Student Workbook

for use with

Welding
Principles and Practices
Third Edition

Raymond J. Sacks

Joseph A. Ciaramitaro, CWE-CWI
Welding Instructor/Industrial Department Chairman
Applied Welding Technology Department
Withlacoochee Technical Institute
Inverness, Florida 34450

Jeffrey Carney, CWI
Department Chair/Associate Professor
Welding Engineering Technology Department
Ferris State University

Boston   Burr Ridge, IL   Dubuque, IA   Madison, WI   New York   San Francisco   St. Louis
Bangkok   Bogotá   Caracas   Kuala Lumpur   Lisbon   London   Madrid   Mexico City
Milan   Montreal   New Delhi   Santiago   Seoul   Singapore   Sydney   Taipei   Toronto

The McGraw·Hill Companies

Student Workbook for use with
WELDING: PRINCIPLES AND PRACTICES, THIRD EDITION
Raymond J. Sacks, Joseph A. Ciaramitaro, and Jeffrey Carney

Published by McGraw-Hill Higher Education, an imprint of The McGraw-Hill Companies, Inc., 1221 Avenue of the Americas, New York, NY 10020. Copyright © 2005 by The McGraw-Hill Companies, Inc. All rights reserved.

No part of this publication may be reproduced or distributed in any form or by any means, or stored in a database or retrieval system, without the prior written consent of The McGraw-Hill Companies, Inc., including, but not limited to, network or other electronic storage or transmission, or broadcast for distance learning.

This book is printed on acid-free paper.

2 3 4 5 6 7 8 9 0 QSR/QSR 0 9 8 7 6 5

ISBN 0-07-825062-5

www.mhhe.com

TABLE OF CONTENTS

Introduction to the Instructor v
Introduction to the Student vii

Chapter 1	History of Welding 1
Chapter 2	Industrial Welding 5
Chapter 3	Steel and Other Metals 9
Chapter 4	Basic Joints and Welds 17
Chapter 5	Gas Welding 21
Chapter 6	Flame Cutting Principles 25
Chapter 7	Flame Cutting Practice 27
Chapter 8	Gas Welding Practice: Jobs 8-J1-J38 31
Chapter 9	Braze Welding and Advanced Gas Welding Practice: Jobs 9-J39-J53 37
Chapter 10	Soldering and Brazing Principles and Practices 43
Chapter 11	Shielded Metal-Arc Welding Principles 49
Chapter 12	Shielded Metal-Arc Welding Electrodes 53
Chapter 13	Shielded Metal-Arc Welding Practice: Jobs 13-J1-J28 (Plate) 59
Chapter 14	Shielded Metal-Arc Welding Practice: Jobs 14-J26-J42 (Plate) 69
Chapter 15	Shielded Metal-Arc Welding Practice: Jobs 15-J43-J55 (Plate) 75
Chapter 16	Shielded Metal-Arc Welding Practice (Pipe) 81
Chapter 17	Arc Cutting Principles and Arc Cutting Practice 91
Chapter 18	Gas Tungsten-Arc and Plasma Arc Welding Principles 95
Chapter 19	Gas Tungsten-Arc Welding Practice (Plate) 101
Chapter 20	Gas Tungsten-Arc Welding Practice (Pipe) 107
Chapter 21	Gas Metal Arc and Flux Cored Arc Welding Principles 111
Chapter 22	Gas Metal-Arc Welding Practice with Solid and Metal Core Wire (Plate) 115
Chapter 23	Flux-Cored Arc Welding Practice (Plate), Submerged Arc Welding, and Related Processes 121
Chapter 24	Gas Metal Arc Welding Practice (Pipe) 127
Chapter 25	High Energy Beams and Related Welding and Cutting Process Principles . . .131
Chapter 26	General Equipment for Welding Shops 133
Chapter 27	Automatic and Robotic Arc Welding Equipment 137
Chapter 28	Joint Design, Testing, and Inspection 141
Chapter 29	Reading Shop Drawings 147
Chapter 30	Welding Symbols 151
Chapter 31	Safety 157
Chapter 32	Welding and Bonding of Plastics 161

INTRODUCTION
To the Instructor

This workbook is designed for use with the textbook *Welding: Principles and Practices, 3rd Ed.* Each exercise has been prepared to assist the student in mastering the information in the text. Every chapter in the workbook has the same number as the textbook chapter it reviews.

Every chapter of the text may not be assigned, and chapters need not be covered in sequence. The workbook items are designed to accommodate whatever approach an instructor may decide. The review questions in this workbook are supplementary to the review questions in the text. These are not intended to be used as replacements for the text's review questions. Also, instructors may wish to develop their own supplementary questions.

A number of approaches are suggested in the assignment of the workbook exercises.

1. The exercises may be completed as "closed book" or "open book" assignments.
2. The material in the textbook may be assigned as homework and the workbook exercises completed at home.
3. The material in the textbook may be assigned as homework and the workbook exercises completed in class.
4. Each student may study the material in the textbook at his own rate of speed and complete the workbook exercises on the same basis.

Raymond J. Sacks   Joseph A. Ciamitaro, CWE- CWI   Jeffrey Carney, CWI

Note: In 1982 and 1998, Joseph A. Ciaramitaro received the prestigious, Howard E. Adkins Memorial Instructors Award as National Teacher of the Year presented by the American Welding Society.

INTRODUCTION
To the Student

Welding is one of the most interesting, challenging and highest paying jobs in industry. It also serves as an important tool in the repair and maintenance field. This workbook includes information on all the forms of welding presented in the textbook, *Welding: Principles and Practices, 3rd Ed.* The exercises are intended to help you achieve as complete an understanding of the fundamentals of welding processes and equipment as possible. The questions cover the information a welder must possess in order to become a skilled craftsman and to gain a place of employment in the welding and advanced fields, such as welding inspector, instructor, and foremen.

The workbook exercises will help you in a number of ways.

1. Surveying the questions will help you determine the information in each chapter to which you should pay special attention.
2. Completing the different kinds of objective questions will be a positive aid in remembering the important aspects of successful welding.
3. Evaluating your success in completing the exercises will indicate not only what you have learned, but also what you need to review.

The following are examples of the types of questions used in the workbook.
All answers should be written in the lined spaces provided on the right side of each page.

1. True or False (True or False). Read the statement carefully and decide if it is true or false. In the space to the right of the question, write *True* if the statement is true or *False* if it is false.

   SAMPLE 1. Welding is an industry that is rapidly going out of existence.
   (True or False)   False

2. Multiple choice with one right answer. These items are unfinished sentences with four or five possible endings, only one of which is correct. Read the statement and the possible conclusions. Select the best answer and place its letter in the answer space at the right.

   SAMPLE 2. St. Louis is on the ____.
   a. Rio Grande.   b. Thames   c. Mississippi   d. Rhine   c

3. Multiple choice with one wrong answer. These items are incomplete statements with four or five possible conclusions, only one of which is wrong. Read the statement and the possible endings. Select the one that is wrong and place its letter in the answer space at the right.

   SAMPLE 3. All of the following are fractions except ____.
   a. 2/3   b. 5/8   c. 6   d. 7/8   e. 9/10.   c

4. **Completion.** Completion items are sentences with blanks where key words should be. Study the sentence carefully and determine which word best completes it. Record this word in the answer space.

   SAMPLE 4. Arc welding is used to join ____.   _____metals_____

5. **Matching.** Matching items are made up of two columns of words or phrases. Those on the left are numbered; those on the right are lettered. Determine the numbered item to which the lettered item is best matched, and record the number in the appropriately lettered answer space to the right of the question.

   SAMPLE 5.
1. River	a. Everest.	a. __2__
2. Mountain	b. Atlantic.	b. __4__
3. Lake	c. Thames.	c. __1__
4. Ocean	d. Ontario.	d. __3__

   Matching items may also call for matching parts of an illustration to corresponding identifying terms. Answers are recorded in the same way as they are for sample question 5.

6. **Identification.** These questions are usually accompanied by illustrations in which names for pieces of equipment or parts of a joint have been replaced by letters. The correct name for a given article should be recorded in the appropriately lettered answer blank.

   SAMPLE 6. Identify the utensils in Sample Fig. 1-1

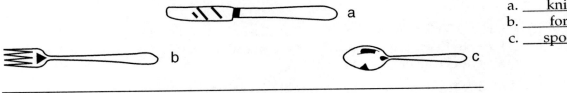

   a. __knife__
   b. __fork__
   c. __spoon__

   SAMPLE FIG. 1-1.

7. **Logical sequence.** Given several steps needed to complete an activity, place them in the order in which they would be undertaken.

NAME: _____   DATE: _____

CHAPTER 1

History of Welding

Choose the letter of the correct answer. If there is no lettered answer, write the word(s) answer.

1. It is always necessary to apply pressure when joining two pieces of metal with the welding process.
   a. True   b. False

2. The melting point of filler metals must be higher than 800°F.
   a. True   b. False

3. The first metal worked by primitive people was ____. (one correct)
   a. iron   b. bronze   c. copper   d. aluminum   e. silver

4. The first welding took place about _____ years ago.

5. The first experiments from which our present welding processes developed began in the early ____. (one correct)
   a. 1600s   b. 1700s   c. 1800s   d. 1900s

6. Welding was an important industrial process long before World War 1.
   a. True   b. False

7. The production demands of World War I introduced welding as a means of product ____. (one correct)
   a. maintenance   b. fabrication   c. repair
   d. stabilization   e. design

8. The first arc welding was performed with a.c. welding machines.
   a. True   b. False

9. The advantages of a.c. welding machines are the absence of arc blow and the high rate of metal deposition.
   a. True   b. False

10. The production demands of World War II hastened the development of ____. (one correct)
    a. bare electrode welding   b. inert gas welding
    c. argon-helium welding   d. metal-arc welding

1. _____

2. _____

3. _____

4. _____

5. _____

6. _____

7. _____

8. _____

9. _____

10. _____

11. Early in the twentieth century, _____ was involved in developing the assembly line method for manufacturing automobiles.

11. _____

12. The development of x-ray of goods called for improved methods of fabrication. Examination of weld metal made it possible to examine the internal soundness of welded joints.
    a. True   b. False

12. _____

13. The drawbacks of inert gas welding were the cost of the _____ and the lack of suitable equipment.

13. _____

14. _____ welding was first used to weld aluminum and magnesium.

14. _____

15. MIG welding was patented before TIG welding.
    a. True   b. False

15. _____

16. MIG welding does not include which one of the following advantages?
    a. small heat affected zone   b. wide bead width
    c. deep penetration   d. faster welding speeds

16. _____

17. Which of these four is not an electric arc process?
    a. atomic-hydrogen welding   b. plasma arc welding
    c. stud welding   d. forge welding
    e. electrogas welding

17. _____

18. Today there are about 25 welding processes in use.
    a. True   b. False

18. _____

19. There are approximately 500,000 welders employed today.
    a. True   b. False

19. _____

20. Women cannot be welders.
    a. True   b. False

20. _____

21. Welding has had very little development in other countries of the world.
    a. True   b. False

21. _____

22. Which of the following is not considered a welding process?
    a. gas   b. resistance   c. brazing   d. soldering   e. arc

22. _____

NAME: _____ DATE: _____

23. Many qualified welders are certified under three of the following four welding codes. Which code should not be included in the list?
a. AWS   b. ASME   c. ADI   d. API

23. _____

24. A diploma from a welding training program is all the certification the bearer needs to perform any welding job.
a. True   b. False

24. _____

25. A welder is rarely called upon to do overhead welding.
a. True   b. False

25. _____

26. Lives may depend upon the quality of a welder's work.
a. True   b. False

26. _____

27. A welder must be able to weld metals other than steel.
a. True   b. False

27. _____

28. In the welding field, there is little opportunity for advancement beyond the job of welder.
a. True   b. False

28. _____

29. In doing a job, each welder takes precautionary measures for only his own safety.
a. True   b. False

29. _____

30. Safety precautions should be followed both in the school shop and in industry.
a. True   b. False

30. _____

Score _____
(30 possible)

NAME: _____ DATE: _____

CHAPTER 2

Industrial Welding

Choose the letter of the correct answer. If there is no lettered answer, write the word(s) answer.

1. The industrial functions of welding are _____ maintenance, and repair.

   1. _____

2. The use of rolled steel in fabrication has the following advantages over the use of castings except one. Which is incorrect?
   a. stronger   b. stiffer   c. more uniform
   d. cheaper   e. rust free

   2. _____

3. High stresses, high temperatures, and high speeds make it difficult to use welding extensively in the space industry.
   a. True   b. False

   3. _____

4. Which of the following is not an advantage of welded construction equipment?
   a. economy   b. size limitation   c. strength
   d. rigidity   e. lightweight

   4. _____

5. The welded joint can be substituted for heavy reinforcing sections in the fabrication of construction equipment.
   a. True   b. False

   5. _____

6. The welded joint is not quite as strong as the sections it joins.
   a. True   b. False

   6. _____

7. Robotic welding is limited to the aerospace industry.
   a. True   b. False

   7. _____

8. Since the mid 1990's, America's road building/repair program has diminished.
   a. True   b. False

   8. _____

9. When fully loaded, a giant off-highway truck can hold _____ cubic yards of earth.

   9. _____

10. Three of the following welding processes play an important part in household equipment. Which one does not?
    a. GTAW   b. GMAW   c. SAW   d. brazing

11. Although steel has many good qualities, cast iron is still the better material for use in machine tools.
    a. True   b. False

12. The production of nuclear energy would not be possible without the use of the welding process.
    a. True   b. False

13. Welded pipe is well suited for use in nuclear plants because it can withstand the high heat and _____ required to operate in such installations.

14. Welded piping systems reduce maintenance costs because their permanently tight connections have greater _____ and rigidity.

15. Welded fittings make it possible to fabricate any combination of sizes and shapes.
    a. True   b. False

16. The use of welding on overland pipelines is limited by their extreme length.
    a. True   b. False

17. Although welding allows more flexibility in the design of freight cars, it also increases a car's weight.
    a. True   b. False

18. A relatively new development in freight-car construction is the super-size, all-welded, aluminum _____.

19. Three manual welding processes are used in the fabrication of railroad equipment. Which of the following is not?
    a. shielded metal arc   b. TIG   c. plasma arc   d. MIG

20. Which of the following is not a major advantage of welded hopper cars?
    a. interior smoothness   b. totally aluminum construction
    c. easy cleaning   d. dust-proof joints

21. Welded passenger coaches are stronger, _____, and more comfortable than riveted coaches.

10. _____

11. _____

12. _____

13. _____

14. _____

15. _____

16. _____

17. _____

18. _____

19. _____

20. _____

21. _____

NAME: _____   DATE: _____

22. Which one of the following four is not an advantage of welded submarines?
    a. reduced size   b. stronger hulls
    c. greater resistance to depth bombs   d. caulked seams

22. _____

23. The welded seams of ships' hulls are difficult to repair.
    a. True   b. False

23. _____

24. Prefabrication, preassembly, and welding were first used in the construction of _____ ships during World War II.

24. _____

25. For over fifty years, bridges have been constructed wholly or in part by the welding process.
    a. True   b. False

25. _____

26. A welded butt joint has less internal stress than a riveted joint.
    a. True   b. False

26. _____

27. In building construction, welding is limited to the structural steel framework and to piping.
    a. True   b. False

27. _____

28. Steel home construction save owners thousands in upkeep, insurance and _____.

28. _____

29. Two serious service failures of riveted tanks were _____ and corrosion around rivets.

29. _____

30. Welding is used extensively in the fabrication of tanks and pressure vessels because of three of the following. Which one is incorrect?
    a. requires less material
    b. develops strength equal to 90% of the tank plate
    c. eliminates seam caulking
    d. increases fabrication speed

30. _____

Score _____
(30 possible)

NAME: _____ DATE: _____

CHAPTER 3

Steel and Other Metals

Choose the letter of the correct answer. If there is no lettered answer, write the word(s) answer.

1. Metals are separated into two major groups, ferrous and ____.

2. Metals that have high iron content are designated as ____ metals.

3. Which of the following is not a nonferrous metal?
   a. nickel
   b. platinum
   c. steel alloys
   d. radioactive metals
   e. aluminum

4. Iron is an alloy of many metals. (True or False)

5. Iron ore is found in abundance only in the U.S. and European countries. (True or False)

6. Iron and ____ are combined to make steel.

7. More welding is performed on ____ than on any other metal.

8. The first people to record the use of iron were the ____.

9. The shaft furnaces used in Europe after A.D. 1350 were the predecessors of modern ____ furnaces.

10. The ____ and cementation processes were used to make steel from ancient times until the development of the Bessemer process.

11. Of the two older steelmaking processes mentioned in question 10, only the ____ process is still in use today.

12. Steelmaking started in the United States about ____ years ago.

1. _____

2. _____

3. _____

4. _____

5. _____

6. _____

7. _____

8. _____

9. _____

10. _____

11. _____

12. _____

CHAPTER 3: Steel and Other Metals | 9

13. Several events spurred the growth of the steel industry. Which of the following did not?
    a. new uses for iron
    b. the shortage of aluminum
    c. the discovery of large iron ore deposits in Michigan
    d. the development of the Bessemer and open hearth process
    e. the expansion of the railroads

13. _____

14. Steel making facilities have not changed over the last few decades. (True or False)

14. _____

15. Japan is the world's largest producer of steel. (True or False)

15. _____

16. The perfection of ____ as a means of joining metals has expanded the use of steel.

16. _____

17. In the United States, nearly all the ore is mined in northern Minnesota near Lake Superior. (True or False)

17. _____

18. The purest iron ore comes from (one right):
    a. United States.   b. India.   c. Brazil.
    d. Sweden.   e. Liberia.

18. _____

19. Iron ore straight from the mine is of suitable quality to feed the blast furnace. (True or False)

19. _____

20. Oxygen is the most abundant element on earth. (True or False)

20. _____

21. Oxygen is used in steelmaking to ____ the product and speed up the process.

21. _____

22. Fuels burned in pure oxygen produce much ____ temperatures than fuels burned in air.

22. _____

23. Heat energy for steelmaking comes from four fuels, three of which are found in nature. Which of the following four does not belong?
    a. coal   b. oil   c. coke   d. natural gas

23. _____

24. Of the four fuels mentioned in question 23, ____ is the most important to the iron and steel industry.

24. _____

25. ____ is made by heating coal to a high temperature in the absence of air.

25. _____

NAME: _____ DATE: _____

26. Nearly 66% of the steel being currently used is recycled. (True or False)

26. _____

27. Scrap steel has become such a valuable commodity that the American Metal Market tracks the price of certain grades of scrap daily. (True or False)

27. _____

28. ____ is used as a flux in the blast furnace.

28. _____

29. Fluxes are used in steelmaking to separate the ____ from the iron ore.

29. _____

30. ____ materials do not melt easily.

30. _____

31. ____ is the residue produced by steelmaking.

31. _____

32. Pure carbon exists in two crystalline forms, diamonds and ____.

32. _____

33. In the ____ furnace, iron is separated from most of the impurities in the ore.

33. _____

34. For making steel, the blast furnace is charged with three of the following. Which is incorrect?
    a. limestone   b. coal   c. iron ore   d. coke

34. _____

35. Liquid iron is poured into molds to make ____ iron.

35. _____

36. The annual production of the product referred to in question 34 has declined because the number of blast furnaces is decreasing. (True or False)

36. _____

37. Melting steel in a ____ purifies the finished product by reducing the amount of unwanted gases in the metal.

37. _____

38. The two processes for vacuum melting are vacuum ____ melting and consumable electrode vacuum arc melting.

38. _____

39. Match the methods of degassing steel to their best descriptions.
    1. Stream
    2. Ladle
    3. Vacuum lifter

    a. Air is removed from a tank containing molten steel.
    b. Molten steel is forced through a nozzle into a vacuum chamber.
    c. Molten steel is poured into a tank from which the air has already been removed.

39a. _____
39b. _____
39c. _____

CHAPTER 3: Steel and Other Metals | 11

40. From the blast furnace, molten steel is poured into ____ molds for cooling.

40. _____

41. Match the steelmaking processes to the phrases which best describe them.

   1. Cementation
   2. Crucible
   3. Bessemer
   4. Open Hearth
   5. Electric furnace
   6. Oxygen
   7. Stora-Kaldo

   a. Liquid slag skimmed off the top of melted, processed steel.
   b. Produces alloy, stainless, and tool steels.
   c. Once the principal method for making U.S. Steel, now accounts for only 3%.
   d. Oldest method of steelmaking.
   e. Main advantage is the short amount of time needed to complete process.
   f. So named because melted steel is accessible through furnace doors for visible inspection and sampling.
   g. Named after one of the oldest continuously operating steel companies in the world.

41a. _____
41b. _____
41c. _____
41d. _____
41e. _____
41f. _____
41g. _____

42. ____ is the process in which molten steel is solidified into a semi-finished billet, bloom, or slab for subsequent finishing.

42. _____

43. Each year steel recycling saves enough energy to electrically power more than ____ households.

43. _____

44. Place the steps in the traditional method of solidifying steel in their proper order.
   a. Immerse ingots in soaking pits.
   b. Transfer ingots to roughing mill.
   c. Pour raw steel into molds.
   d. Send blooms or billets to finishing mills.
   e. Remove solidified steel from molds.

44. 1. _____
    2. _____
    3. _____
    4. _____
    5. _____

12 | STUDENT WORKBOOK

NAME: _____ DATE: _____

45. Match the products that result from steel deoxidation to their best descriptions.
    1. Killed
    2. Semikilled
    3. Capped
    4. Rimmed

    a. Very ductile, low-carbon surface layer.
    b. High degree of composition and property uniformity.
    c. Possibly uneven carbon distribution but fairly uniform composition.
    d. Think, low-carbon rim, but otherwise high degree of uniformity.

    45a. _____
    45b. _____
    45c. _____
    45d. _____

46. Refining the grain of steel (one wrong):
    a. strengthens it.
    b. improves ductility.
    c. reinforces the case structure.
    d. increases shock resistance.

    46. _____

47. Using a steam hammer to reduce cast steel to the desired shape is called ____.

    47. _____

48. Rolled steel ingots are called blooms, billets, or ____.

    48. _____

49. Match the terms of various types of flat-rolled steel to their best descriptions.
    1. Black iron
    2. Galvanized
    3. Tin plate
    4. Terne plate

    a. Coated with zinc.
    b. Coated with lead and tin.
    c. Untreated, hot rolled.
    d. Coated with tin.

    49a. _____
    49b. _____
    49c. _____
    49d. _____

50. Tubular steel products are referred to as seamless or ____, depending upon how they are manufactured.

    50. _____

51. An extrusion is formed by drawing metal through an opening. (True or False)

    51. _____

52. When metal is hammered, rolled, or drawn at ordinary temperatures, it is called ____ working.

    52. _____

53. Heating and cooling a metal to improve its physical or structural properties is called ____ treatment.

    53. _____

CHAPTER 3: Steel and Other Metals | 13

54. Match the following heat treatments to the phrases which best describe them.
   1. Hardening   a. Alters ductility, toughness, or electrical or magnetic properties.
   2. Case hardening   b. Produces wear-resistant surface with soft, tough interior.
   3. Annealing   c. Reduces hardness after heat treatment and relieves stresses caused by quenching.
   4. Tempering   d. Improves grain structure after welding, casting, or forging.
   5. Normalizing   e. Increases hardness of medium to very high carbon steel.

   54a. _____
   54b. _____
   54c. _____
   54d. _____
   54e. _____

55. Match the terms used to describe various physical properties of metals to their definitions.
   1. Melting Point   a. Ease with which a metal is vaporized.
   2. Fusibility   b. Brittleness of hot metal.
   3. Volatility   c. Temperature at which a solid becomes liquid.
   4. Thermal conductivity   d. Degree to which metal contains slag, inclusions, and gas pockets.
   5. Hot shortness   e. Ability to change shape without breaking.
   6. Porosity   f. Ease with which a metal may be melted.
   7. Plasticity   g. Ability to carry heat.

   55a. _____
   55b. _____
   55c. _____
   55d. _____
   55e. _____
   55f. _____
   55g. _____

56. Desirable qualities in metals include all of the following except one. Which one is incorrect?
   a. strength   b. toughness   c. shock resistance
   d. brittleness   e. hardness

   56. _____

57. The science that deals with the internal structure of metals is called ____.

   57. _____

58. Failure of metals under repeated or alternating stresses is known as ____ failure.

   58. _____

59. ____ loading is another way of referring to fatigue testing.

   59. _____

60. Aluminum has a higher melting point than lead. (True or False)

   60. _____

NAME: _____   DATE: _____

61. An increase in carbon content in steel makes the steel softer. (True or False)

61. _____

62. All but one of the following are among the desirable qualities of copper. Which one is not?
    a. malleability   b. ductility   c. corrosion resistance
    d. hardness   e. thermal conductivity

62. _____

63. Sulfur improves the quality of steel. (True or False)

63. _____

64. Aluminum is found in abundance in the pure state. (True or False)

64. _____

65. All but one of the following metals increase the corrosion resistance of steel. Which one does not?
    a. manganese   b. molybdenum   c. chromium
    d. nickel   e. columbium

65. _____

66. High carbon steels have a carbon content of approximately 0.30 to 0.60 per cent. (True or False)

66. _____

67. The metal most commonly added to iron and copper to improve the corrosion resistance of stainless steel is ____.

67. _____

68. Tool steels have low carbon content. (True or False)

68. _____

69. The four types of cast iron are listed below. Which one in the list is wrong?
    a. gray   b. malleable   c. ductile   d. white   e. nodular

69. _____

70. The two major aspects of contraction in all types of welding are distortion and ____.

70. _____

71. Steps that can be taken before welding to prevent distortion are listed below. Which is incorrect?
    a. arranging joints so they balance each other
    b. placing parts in the exact position they should occupy when the weld is completed
    c. prebending parts to be welded
    d. using semi-automatic, submerged arc welding processes

71. _____

72. Welding both sides of a joint at the same time virtually eliminates distortion. (True or False)

72. _____

73. All metals expand at the same rate. (True or False)

73. _____

CHAPTER 3: Steel and Other Metals | 15

74. Match the methods of controlling residual stress to their best descriptions,

1. Postheating
2. Full annealing
3. Cold peening
4. Preheating

a. Controls expansion and contraction during the welding operation.
b. Most effective method, but difficult to control.
c. Most commonly used method.
d. Stretches bead by hammering.

74a. _____

74b. _____

74c. _____

74d. _____

75. Using a low frequency, high amplitude vibration to reduce residual stress levels to a point where they cannot cause distortion or other problems is known as ____ stress relieving.

75. _____

Score _____
(100 possible)

NAME: _____   DATE: _____

CHAPTER 4

Basic Joints and Welds

Choose the letter of the correct answer. If there is no lettered answer, write the word(s) answer.

1. Identify the five basic types of joints shown in Fig. 4-1.

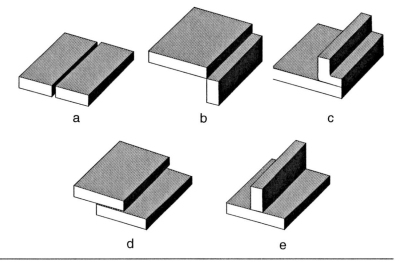

FIG. 4-1.

1a. _____

1b. _____

1c. _____

1d. _____

1e. _____

2. The welder need not be concerned about design and welding procedures. This is the responsibility of the engineer. (True or False)

2. _____

3. There are only four basic types of welds. Which one of the following is not?
   a. bead   b. groove   c. beam   d. fillet   e. plug

3. _____

4. Open corner joints are also welded with fillet welds. (True or False)

4. _____

5. Groove welds are used for butt joints. (True or False)

5. _____

6. Strength welds are usually designed to possess the maximum ____ characteristics of the base metal.

6. _____

7. Strength welds may be all but one of the following. Which one is incorrect?
   a. bead   b. edge   c. groove   d. fillet

7. _____

CHAPTER 4: Basic Joints and Welds | 17

8. A groove weld is measured and sized by its depth of ____ fusion into the joint.

9. High reinforcement increases the strength of the welded joint. (True or False)

10. Proper reinforcement should not exceed ____ inch.

11. Which one of the following is not a part of a 3/4 butt weld:
    a. throat   b. heel   c. shoulder   d. overlap   e. face

12. The most common weld used in industry is the ____ weld.

13. Contour is the shape of the ____ of the weld.

14. Fillet welds can have three types of contours. Which of the following is incorrect?
    a. context   b. concave   c. flat   d. convex

15. Three different throat sizes may be referred to when discussing the size of fillet welds. Which one of the following is incorrect?
    a. theoretical   b. effective   c. acute   d. actual

16. Fillet and groove welds are usually made along the ____ length of the joint.

17. Match the types of welds to their best descriptions.
    1. Strength   a. Extends the entire length of a joint.
    2. Caulking   b. Makes riveted joints leak proof.
    3. Composite   c. Holds job parts together in preparation for welding.
    4. Continuous   d. Carries the structural load.
    5. Intermittent   e. Placed at intervals to reduce cost of non-critical work.
    6. Tack   f. Meets load requirements and is also leak proof.

8. _____

9. _____

10. _____

11. _____

12. _____

13. _____

14. _____

15. _____

16. _____

17a. _____

17b. _____

17c. _____

17d. _____

17e. _____

17f. _____

NAME: _____ DATE: _____

18. Match the welding positions to their descriptions.
   1. Flat
   2. Horizontal
   3. Vertical
   4. Overhead

   a. Weld travel may be up or down.
   b. Filler deposited from upper side of joint: weld face is horizontal.
   c. Filler metal deposited from underside of joint; weld face is horizontal.
   d. Filler is deposited on upper side of horizontal surface against a vertical surface.

   18a. _____
   18b. _____
   18c. _____
   18d. _____

19. Metal extending above the surfaces of the plate is called ____.

   19. _____

20. High weld reinforcement increases the strength of a weld. (True or False)

   20. _____

21. The width of a groove weld should not extend beyond the shoulder of the joint more than ____.
   a. 3/8"   b. 1/2"   c. 1/16"   d. 1/8"   e. 1/4"

   21. _____

22. The distance from the root to the toe of a fillet weld is called the ____.

   22. _____

23. Throat thickness is the distance from the root to the ____ of a fillet weld.

   23. _____

24. The ideal fillet weld has a ____ or slightly convex face and equal leg length.

   24. _____

25. A concave fillet weld is stronger and more economical than other fillet welds. (True or False)

   25. _____

26. Generally, welded joints are at least as strong as the base metal being welded. (True or False)

   26. _____

27. The strength of a joint depends upon all but one of the following. Which one is incorrect?
   a. strength of the weld metal
   b. type of joint penetration
   c. heat treatment
   d. position of welding
   e. location of joint in the assembly

   27. _____

28. A weld ____ is any interruption in normal flow of the structure of a weldment.

   28. _____

29. Match the weld defects to their best descriptions.

1. Insufficient throat	a. Incomplete fusion in a fillet weld, leaving no root penetration.	29a. _____
2. Excessive convexity	b. An overflow of weld metal beyond the end of fusion.	29b. _____
3. Undercut	c. Slag and gas entrapped in the filler metal.	29c. _____
4. Overlap	d. Fillet weld weakened by shortening a leg.	29d. _____
5. Insufficient leg	e. Throat thickness less than plate thickness.	29e. _____
6. Bridging	f. Opposite of a concave profile.	29f. _____
7. Porosity	g. A cutting away of plate surface at the weld toe reduces plate thickness.	29g. _____

Score _____
(47 possible)

NAME: _____   DATE: _____

CHAPTER 5

Gas Welding

Choose the letter of the correct answer. If there is no lettered answer, write the word(s) answer.

1. The oxyacetylene process is used for all but one of the following operations. Which one is incorrect?
   a. brazing b. soldering c. metalizing
   d. large diameter pipe welding
   e. general maintenance and repair

   1. _____

2. The fuel gases commonly used for gas welding include all but one of the following. Which one is incorrect?
   a. acetylene b. Mapp gas c. propane
   d. coal gas e. natural gas

   2. _____

3. Oxygen burns at a very high temperature. (True or False)

   3. _____

4. Oxygen may be manufactured by the liquefaction of air or the ____ of water.

   4. _____

5. To conserve oxygen, open cylinder valves only part way. (True or False)

   5. _____

6. Oxygen cylinder valves should be greased regularly so that they operate properly. (True or False)

   6. _____

7. Because of the pressure involved, oxygen cylinders undergo rigid testing and inspection. (True or False)

   7. _____

8. Acetylene is generated as the result of a chemical reaction between calcium carbide and ____.

   8. _____

9. Because of the danger of storing highly pressurized acetylene, cylinders are filled with ____.

   9. _____

10. For safety purposes, open acetylene cylinder valves only partway. (True or False)

    10. _____

11. Match the fuel gases to the temperature of their flames under neutral conditions.
    1. Acetylene   a. 4,600° F.
    2. Mapp gas   b. 5,301° F.
    3. Propane   c. 5,190° F.
    4. Natural gas   d. 5,420° F.
    5. Hydrogen   e. 5,000° F.

    11a. _____
    11b. _____
    11c. _____
    11d. _____
    11e. _____

12. An effective welding fuel gas must possess all but one of the following characteristics. Which one is incorrect?
    a. high flame temperature
    b. high rate of flame propagation
    c. sharp preheat cones
    d. adequate heat
    e. minimum chemical reaction of the flame with vase and filler metal

    12. _____

13. An acetylene cylinder may be stored lying on its side. (True or False)

    13. _____

14. Mapp gas is the safest industrial fuel. (True or False)

    14. _____

15. All portable cylinders used for the storage and shipment of compressed gases shall be constructed and maintained in accordance with the regulations of the ____.

    15. _____

16. Approved protective equipment shall be installed in OAW systems to prevent all but one of the following. Which one is incorrect?
    a. backflow of oxygen into fuel-gas supply system
    b. over acetylene gas pressure
    c. passage of a flash back into the fuel-gas supply system
    d. excessive back pressure of oxygen in the fuel-gas supply system

    16. _____

17. Increasing the temperature of a cylinder of gas will ____ the pressure within the tank.

    17. _____

18. Use a gas ____ to avoid frequently relighting the torch and adjusting the flame.

    18. _____

22 | STUDENT WORKBOOK

NAME: _____ DATE: _____

19. Match the pieces of equipment to their functions.
   1. Torch   a. Ignites the oxyacetylene mixture.   19a. _____
   2. Regulator   b. Carries the oxygen and acetylene cylinders.   19b. _____
   3. Hose   c. Used for cutting.   19c. _____
   4. Goggles   d. Reduces tank pressure to torch pressure.   19d. _____
   5. Lighter   e. Carries gas from tank to torch.   19e. _____
   6. Truck   f. Required for safety.   19f. _____

20. Single-stage regulators are for use with ____ systems.   20. _____

21. The two types of torches are the balanced pressure and the ____.   21. _____

22. Use torch ____ to produce a variety of flame sizes.   22. _____

23. Most welding tips are made of pure drawn ____.   23. _____

24. Match the hose colors to the gases they conduct.
   1. Green   a. Acetylene   24a. _____
   2. Red   b. Air   24b. _____
   3. Black   c. Oxygen   24c. _____

25. Oxygen-hose couplings have ____ hand threads.   25. _____

26. A ____ is a cleaning agent used to dissolve oxides and cleanse metals for welding, soldering, and brazing.   26. _____

27. Hydrogen welding is used primarily for welding metals with ____ melting points.   27. _____

28. A special regulator is needed when using propane, butane, city gas, or natural gas as a fuel gas for welding. (True or False)   28. _____

29. ____ gas is used extensively in the plumbing, refrigeration, and electrical trades for heating and soldering operations.   29. _____

30. Acetylene burned with air produces lower flame temperatures than the other gas combinations. (True or False)   30. _____

CHAPTER 5: Gas Welding | 23

31. Carbon in the form of rods, plates, or paste is used for all but one of the following. Which one is incorrect?
    a. to protect surfaces and holes
    b. to make back-up welds
    c. to supply additional carbon to the molten metal
    d. to control and shape the flow of metal
    e. to support and align broken parts

31. _____

32. Eyeglass-types frames are recommended for welding lenses to be worn indoors because they are light and comfortable to wear. (True or False)

32. _____

33. Match the welding lens shade numbers to the type of work for which they are recommended.
    1. Shade 2   a. Light brazing.
    2. Shade 3   b. Standard acetylene gas welding.
    3. Shade 4   c. Acetylene brazing.
    4. Shade 5   d. Carbon arc cutting and welding.
    5. Shade 6   e. Electric arc welding between 75 and 250 amperes.
    6. Shade 8   f. Low-amperage electric arc welding.
    7. Shade 10   g. Electric arc cutting above 250 amperes.
    8. Shade 12   h. Light acetylene cutting.
    9. Shade 14   i. Low temperature furnace work.

33a. _____
33b. _____
33c. _____
33d. _____
33e. _____
33f. _____
33g. _____
33h. _____
33i. _____

Score _____
(52 possible)

NAME: _____ DATE: _____

CHAPTER 6

Flame Cutting Principles

Choose the letter of the correct answer. If there is no lettered answer, write the word(s) answer.

1. Oxyacetylene cutting is excellent for cutting nonferrous metals. (True or False)

2. Oxyacetylene cutting is a very rapid form of rusting. (True or False)

3. The oxy-fuel gas cutting process uses only acetylene as a fuel gas. (True or False)

4. An adaptable cutting attachment makes it possible to use a welding torch as a cutting tool without disconnecting the hose. (True or False)

5. The central hole in a cutting tip provides for the passage of ___.

6. The holes around the center hole of a cutting tip are ___ flame holes.

7. A cutting torch should be ignited only with a friction lighter or safety matches. (True or False)

8. Both goggles and ___ must be worn during the cutting process for protection from sparks, metal particles, and heat.

9. To cut straight lines accurately, you use which one of the following?
    a. straight gouging squares b. magnetic burning tips
    c. rivet-burning tips d. slicing gauging tips
    e. magnetic burning squares

10. Oxy-fuel gas cutting machines are able to cut only a limited number of designs. (True or False)

11. Cutting tips are designated as standard or ___.

12. The various fuel gases require different volumes of oxygen and ___.

1. _____
2. _____
3. _____
4. _____
5. _____
6. _____
7. _____
8. _____
9. _____
10. _____
11. _____
12. _____

13. It is possible to cut geometric designs into metal sheets without using templates. (True or False)

13. _____

14. For plates up to 1/2-inch thick, ____ cutting is an efficient method of cutting several thicknesses of material at once.

14. _____

15. A portable ____ cutter may be used for trimming and beveling angles and channels.

15. _____

16. A length of black iron pipe attached to a fuel source and used for cutting heavy sections of steel is an oxygen ____ cutter.

16. _____

17. The cutter in question 14 is especially useful for all but one of the following. Which one is incorrect?
    a. piercing holes in heavy, thick steel
    b. beveling heavy pipe
    c. cutting off large risers
    d. opening plugged holes in steelmaking equipment

17. _____

Score _____
(17 possible)

NAME: _____ DATE: _____

CHAPTER 7

Flame Cutting Practice

Choose the letter of the correct answer. If there is no lettered answer, write the word(s) answer.

1. Flame cutting is an electric welding process. (True or False)

2. Flame-cutting processes depend upon the fact that all metals ____ to a certain degree.

3. The cutting torch provides the heating flame, maintains the temperature, and directs a ____ stream on the cutting point.

4. Nearly all flame-cutting problems are caused by obstruction of the nozzle tip. (True or False)

5. Mechanical cutting may damage the plate edge of the cut metal. (True or False)

6. The flame-cut edge of high carbon steel has a tendency to harden and crack. (True or False)

7. The function of the fuel gas is to feed the ____ flames.

8. Which of the following fuel gases is the most commonly used?
   a. propane   b. Mapp gas   c. acetogen
   d. acetylene   e. natural gas

9. The two gases preferred for underwater cutting are ____.
   a. acetylene and acetogen   b. natural gas and hydrogen
   c. Mapp gas and propane   d. acetylene and hydrogen

10. The hottest flame is produced by ____.
    a. acetylene   b. propane   c. natural gas
    d. hydrogen   e. acetogen

11. Mapp gas produces more heat, measured in Btus, than acetylene. (True or False)

1. _____

2. _____

3. _____

4. _____

5. _____

6. _____

7. _____

8. _____

9. _____

10. _____

11. _____

12. The fuel gas that concentrates the most Btus in one area is ____.
    a. acetylene   b. propane   c. natural gas
    d. hydrogen   e. Mapp gas

13. Acetylene is expensive to use because it requires more oxygen for the cutting process than other gases do. (True or False)

14. The most economical fuel gas, widely used in steel mills for removing surface defects, is ____.
    a. acetylene   b. propane   c. natural gas
    d. hydrogen   e. Mapp gas

15. For cutting plate six or more inches thick, the better choices are Mapp gas or ____.

16. Oxyacetylene cutting is most commonly used on ____ and low alloy steels.

17. The rate of oxidation decreases as the carbon content of metals increases. (True or False)

18. All metal oxides melt at a lower temperature than the base metal. (True or False)

19. Match the methods for speeding the cutting process to their best descriptions.
    1. Preheating   a. Melting a low carbon steel plate forms, an excess of iron oxide, allowing a continuous cut.
    2. Waster plate   b. Moving the torch from side to side oxidizes additional material.
    3. Wire feed   c. Increasing the temperature of the material to be cut increases the rate of oxidation.
    4. Oscillatory motion   d. Burning low carbon steel wire brings the surface of a metal plate to ignition temperature.

20. The gap created as material is removed by cutting is called the ____.

12. _____

13. _____

14. _____

15. _____

16. _____

17. _____

18. _____

19. _____

20. _____

NAME: _____   DATE: _____

21. The width of the gap produced by the cut increases as the thickness of the material increases. (True or False)

21. _____

22. The flow of high-pressure oxygen may form ____ lines on the face of the work.

22. _____

23. Match the types of cutting to their best descriptions.
   1. Straight line cutting   a. Forms a hole in a metal part.
   2. Bevel cutting   b. Torch is held perpendicular to plate.
   3. Flame piercing   c. Removes a narrow strip of surface metal without penetrating the plate.
   4. Flame scarfing   d. Removes a threaded bolt from a threaded hole without destroying hole or threads.
   5. Flame washing   e. Remove cracks, surface seams, scabs, and other defects from unfinished steel.
   6. Flame gouging   f. Torch tip is held sideways at a designated angle.

23a. _____
23b. _____
23c. _____
23d. _____
23e. _____
23f. _____

24. One of the principle uses for oxyacetylene flame cutting is the preparation of plate and ____ for welded fabrications.

24. _____

25. Holding the nozzle tip too close to the plate will cause ____ at the top edge.

25. _____

26. Unsteady torch travel produces a wavy, ____ cut.

26. _____

27. Preheat flames that are too ____ will cause too much slag.

27. _____

28. Inadequate preheat with flames held too far from the plate produces a ____ too wide at the top.

28. _____

29. To increase cutting speed for thicker materials, it is necessary to ____ tip size.

29. _____

30. The working pressure on the regulators is set by turning the adjusting screw to the ____.

30. _____

31. A carburizing flame contains an excessive amount of ____.

31. _____

CHAPTER 1: History of Welding | 29

32. As the oxygen valve is adjusted so that the secondary cone of the carburizing flame disappears, the ____ flame is formed.

32. _____

33. If a cut surface is to be used for welding, a ____ flame is recommended.

33. _____

34. The fastest preheat time is achieved with a reducing flame. (True or False)

34. _____

35. For a cutting operation, the preheat flames should be in contact with the metal. (True or False)

35. _____

36. A cut that has been started properly will produce a shower of ____ on the underside of the plate.

36. _____

37. During a satisfactory cut, slag will flow freely from the ____.

37. _____

38. The deposit resulting from the oxygen cutting process, which adheres to the base metal or cut surface, is known as ____.

38. _____

39. To make cuts in cast iron, use a carburizing flame and a ____ motion of the torch.

39. _____

40. The preheat flames for cutting cast iron must be smaller than those used for steel. (True or False)

40. _____

41. The oxygen pressure for cutting cast iron must be greater than that for steel of the same thickness. (True or False)

41. _____

42. When working with low-grade cast iron, feeding a steel rod into the kerf to increase slag fluidity is a form of ____ cutting.

42. _____

43. There is always considerable slag when cast iron is cut. (True or False)

43. _____

Score _____
(43 possible)

NAME: _____   DATE: _____

CHAPTER 8

Gas Welding Practice: Jobs 8-J1-J38

Choose the letter of the correct answer. If there is no lettered answer, write the word(s) answer.

1. Match following terms to their best definitions.
   1. Fusion
   2. Adhesion
   3. Chamfering
   4. Penetration
   5. Reinforcement

   a. Grooving.
   b. Building weld metal above the surface of the base metal.
   c. Failure of base and filler metals to join completely.
   d. Depth to which base metal is melted and joined.
   e. The essential characteristic of a good weld.

   1a. _____
   1b. _____
   1c. _____
   1d. _____
   1e. _____

2. A neutral flame does all but one of the following. Which one is incorrect?
   a. causes no chemical change in the weld metal
   b. produces the highest temperature of the three oxyacetylene flames
   c. serves as the reference point for flame adjustments
   d. is used for most oxyacetylene cutting and welding operations

   2. _____

3. Three of the following four are true concerning an excess acetylene flame. Which one is incorrect?
   a. introduces carbon into the weld pool
   b. is a reducing flame
   c. creates a cloudy, boiling weld pool when used on steel
   d. produces a ductile but tough weld

   3. _____

4. An excess oxygen flame does all but one of the following. Which one is incorrect?
   a. removes excess oxygen from iron oxides when used on steel
   b. reduces flame temperature with too great an oxygen increase
   c. produces a flame hotter than the carburizing flame
   d. seriously reduces weld quality when used with oxidizing metals

   4. _____

5. A carburizing flame is recommended for welding all but one of the following. Which one is incorrect?
   a. high-test pipe   b. Wrought iron
   c. chrome   d. alloy steels

5. _____

6. An oxidizing flame is recommended for welding all but one of the following. Which one is incorrect?
   a. sheet brass   b. bronze
   c. chromium nickel   d. steel plate

6. _____

7. Maintenance welding requires the use of ____ welding equipment.

7. _____

8. An oxygen hose may be green or black. (True or False)

8. _____

9. An acetylene hose may be red or white. (True or False)

9. _____

10. All right-hand threads are ____ connections.

10. _____

11. High gas pressure is often the major cause of poor ____.

11. _____

12. When closing down equipment, turn off the ____ valve on the torch first.

12. _____

13. When closing down portable or line equipment, always release the regulator adjusting screws. (True or False)

13. _____

14. After the cylinder valves are closed, torch valves are re-opened to drain gas lines. (True or False)

14. _____

15. Neither check valves nor flashback arrestors will protect the torch or tip. (True or False)

15. _____

16. Information on safety operation procedures can be obtained by contacting the ____.

16. _____

17. An important factor in making a ripple weld is the ____ speed of the flame.

17. _____

18. Moving the welding flame across the plate too fast causes all but one of the following. Which one is incorrect?
    a. loss of weld pool   b. poor fusion
    c. burn-through   d. a narrow bead

18. _____

32 | STUDENT WORKBOOK

NAME: _____ DATE: _____

19. A satisfactory ripple weld has even ripples of uniform width. (True or False)

19. _____

20. To form the molten pool for ripple welding, hold the inner core of the torch flame ____ inch above the plate surface.

20. _____

21. For ripple welding in the vertical position, hold the torch like a pencil. (True or False)

21. _____

22. Edge and corner joints may be welded with or without filler rod. (True or False)

22. _____

23. If the filler rod is permitted to drip from above the weld pool, the molten drop will become ___.

23. _____

24. In bead welding, the cone of the flame should touch the molten weld pool. (True or False)

24. _____

25. If the filler rod is of the proper size it should not melt in the ____.

25. _____

26. The ____ welding position is often the most difficult to master because of the weld's tendency to sag.

26. _____

27. To be sure of maximum strength for welded edge or corner joints, filler rod must be used. (True or False)

27. _____

28. Corner-joint penetration should show through to the backside of the weld. (True or False)

28. _____

29. Use the torch flame to protect the weld pool from the formation of oxides and ____.

29. _____

30. Square butt joints require no edge preparation for welding. (True or False)

30. _____

31. In making a butt weld, melt the plate ____ before adding filler rod.

31. _____

32. Butt-weld reinforcement should be no higher than which one of the following?
    a. 1/32 inch   b. 3/32 inch
    c. 1/6 inch   d. 1/8 inch

32. _____

33. Bending a good butt weld will produce no cracks or breaks. (True or False)

33. _____

34. A lap joint is welded with a ____ weld.

35. Maximum lap-joint strength requires welding on both sides. (True or False)

36. It is possible for the top sheet to burn away during lap welding while the bottom sheet remains too cool. (True or False)

37. During lap welding, protect the top plate from excess heat with which one of the following?
    a. filler rod   b. torch position
    c. the Weld pool   d. tack welds

38. The difference between sheet steel and steel plate is the ____ of the metal.

39. When welding a T-joint, all but one of the following is correct. Which one is incorrect?
    a. the vertical plate tends to melt first
    b. use the filler rod to help prevent undercut
    c. use less heat than you would for other joints
    d. form a slightly convex weld face

40. When welding the second side of a T-joint, direct more heat toward the flat plate. (True or False)

41. Light gauge sheet metal should be welded with a ____ type-welding torch.

42. Pipe and heavy plate are welded with ____ test filler rod.

43. For ____ plate welding, the travel is from left to right and the flame is directed backward.

44. The surface appearance of a backhand weld is smoother than that of a forehand weld. (True or False)

45. Because of the constantly changing torch and rod angles required ____ welding is recommended for a welder's first pipe practice.

46. For pipe welding, use one weld pass for each 1/8-inch thickness of pipe wall over 3/16 inch. (True or False)

34. _____

35. _____

36. _____

37. _____

38. _____

39. _____

40. _____

41. _____

42. _____

43. _____

44. _____

45. _____

46. _____

NAME: _____   DATE: _____

47. For forehand welding in the horizontal fixed position, travel begins at the ____ of the pipe.

47. _____

48. Backhand pipe welding is based on the principle that hot steel will absorb ____, which lowers the melting.

48. _____

49. When the backhand technique is used, vertical pipe welding can travel from top to bottom. (True or False)

49. _____

Score _____
(53 possible)

NAME: _____ DATE: _____

CHAPTER 9

Braze Welding and Advanced Gas Welding Practice: Jobs 9-J39-J53

Choose the letter of the correct answer. If there is no lettered answer, write the word(s) answer.

1. For braze welding all but one of the following is true. Which one is incorrect?
   a. is done with a bronze filler rod
   b. does not require fusion of the base metal
   c. requires a special joint design to be used only for torch brazing
   d. joins dissimilar metals

   1. _____

2. For braze welding all but one of the following is true. Which one is incorrect?
   a. is recommended for parts to be used in high temperatures
   b. reduces expansion and contraction
   c. increases welding speed
   d. is the only means of welding malleable cast iron

   2. _____

3. Bronze filler rod melts at ____ degrees F.

   3. _____

4. Allowing a small amount of molten bronze filler rod to spread over a joint is called ____.

   4. _____

5. If the base metal produces white smoke during braze welding, the surface is too ____.

   5. _____

6. For the graphite in gray cast iron all but one of the following is true. Which one is incorrect?
   a. makes machining easy
   b. lessens tensile strength
   c. limits welding to the fusion method
   d. causes low ductility

   6. _____

7. Gray cast iron is ____ welded when the color of the base metal must be retained.

   7. _____

8. When fusion welding cast iron, stress caused by expansion and contraction may be reduced by ____.

9. Blowholes and porosity in cast iron welds may be caused by the use of too much ____.

10. For aluminum all but one of the following is true. Which one is incorrect?
    a. is lightweight but strong
    b. is best suited to oxyacetylene welding
    c. conducts electricity well
    d. resists corrosion
    e. does not change color when heated

11. Aluminum melts at ____ temperature than copper and steel.

12. Aluminum's tendency to collapse when it reaches its melting point is called ____.

13. Aluminum welds have a tendency to ____ because of their high expansion and contraction rate.

14. The diameter of an aluminum filler rod should equal the ____ metal being welded.

15. To save time, flux may be applied to filler rods well in advance of welding. (True or False)

16. For powdered aluminum-welding flux all but one of the following is true. Which one is incorrect?
    a. may be mixed with alcohol or water
    b. should coat the welding rod used on sheet aluminum
    c. is applied to the end of filler rod used for cast aluminum
    d. must be mixed in steel containers to avoid contamination

17. A soft flame is essential in all oxy-gas welding of aluminum. (True or False)

18. All aluminum joints require edge preparation before welding. (True or False)

19. Aluminum welding may be done in all positions, but the ____ position is preferred.

8. _____

9. _____

10. _____

11. _____

12. _____

13. _____

14. _____

15. _____

16. _____

17. _____

18. _____

19. _____

NAME: _____ DATE: _____

20. Flux residues on gas-welded sections may ____ aluminum.

20. _____

21. Aluminum welds may be steam cleaned or dipped in a ____ bath.

21. _____

22. Stirring the filler rod in the aluminum weld pool reduces porosity by bringing ____ to the surface of the pool.

22. _____

23. The cone of the flame should touch the weld pool to assist in moving it along. (True or False)

23. _____

24. Weld cracking is a greater problem in welding sheet aluminum than cast aluminum. (True or False)

24. _____

25. Aluminum oxide has a higher melting point than cast aluminum. (True or False)

25. _____

26. In the term 18-8 stainless steel, the 18 indicates the amount of ____ present.

26. _____

27. When oxyacetylene welding stainless steel, three of the following are true. Which one is incorrect?
    a. use a larger torch tip than for carbon steel
    b. a flux is required to protect the weld surface
    c. some chromium is lost because of oxidation
    d. a precise neutral flame must be maintained

27. _____

28. Preheating stainless steel prevents air hardening and cracking. (True or False)

28. _____

29. When welding stainless steel, the filler rod should be held within the flame to prevent oxidation. (True or False)

29. _____

30. The scale on a stainless steel weld that does not need to be polished may be removed with acid or by ____.

30. _____

31. Oxyacetylene fusion welding of copper may be used only on ____ copper base metals.

31. _____

32. When oxyacetylene welding is applied to copper three of the following are true. Which one is incorrect?
    a. joint design is basically the same as that used for steel
    b. flux is not generally required
    c. the filler rod should melt at a slightly higher temperature than the base metal
    d. use a larger tip than for the same thickness of steel

32. _____

33. For copper three of the following are true. Which one is incorrect?
    a. has high heat conductivity
    b. can be hardened by heat treatment
    c. melts immediately upon reaching 1,980 degrees F.
    d. forms a highly fluid weld puddle

33. _____

34. For nickel three of the following are true. Which one is incorrect?
    a. is very strong
    b. resists corrosion
    c. withstands high temperature
    d. is suited to heat treatment
    e. resists contamination and can be welded without cleaning

34. _____

35. For oxyacetylene welding of nickel, a ____ degree V is required.

35. _____

36. The nickel alloy with greater tensile strength than steel is annealed ____.

36. _____

37. For magnesium all but one of the following is true. Which one is incorrect?
    a. is lighter than aluminum
    b. is usually alloyed with aluminum
    c. is strongest used in its pure state
    d. resists corrosion

37. _____

38. When welding magnesium with the oxyacetylene process all but one of the following is correct. Which one is incorrect?
    a. avoid lap and fillet welds
    b. support joints to prevent collapse
    c. use flux sparingly
    d. preheat for easier removal of oxide film
    e. treat finished weld with sodium dichromate

38. _____

39. For lead three of the following are true. Which one is incorrect?
    a. is malleable and ductile
    b. is combined with antimony for strength
    c. conducts electricity
    d. withstands corrosive liquids

39. _____

NAME: _____ DATE: _____

40. Problems encountered when oxyacetylene welding lead include three of the following. Which is not true?
    a. edge preparation
    b. proper torch technique
    c. excess fusion
    d. burn-through

40. _____

41. For hard facing a metal surface exposed to much wear three of the following are true. Which one is incorrect?
    a. adds no new qualities to the base metal
    b. increases service life as much as 40 times
    c. allows rebuilding rather than replacement
    d. allows the use of cheaper materials for the part being surfaced

41. _____

42. Match the hard-facing welding processes to the results they achieve.
    1. Oxyacetylene   a. Flawless surface.
    2. Gas tungsten-arc   b. Light surface coating on heavy sections.
    3. Shielded metal-arc c. Suitable surface for most applications.
    4. Atomic hydrogen   d. Heavy metal deposit.

42. _____

43. Three of the following hard-surfacing rods are used to produce resistance to abrasion. Which one is not?
    a. cobalt baser
    b. iron base
    c. copper base
    d. tungsten carbide

43. _____

44. When applying hard-facing materials to steel, all but one of the following are correct. Which one is incorrect?
    a. preheat
    b. use an oxidizing flame to produce sweating
    c. allow the rod to lightly touch the sweating area
    d. use the flame's pressure to move the weld puddle
    e. cool slowly

44. _____

45. Match the following materials to the special handling they require for hard surfacing.
    1. Cast iron   a. Use less acetylene in the torch flame.
    2. Alloy steels   b. Anneal fully and cool very slowly.
    3. High speed steels   c. Preheat and post cool carefully to avoid cracking.

45a. _____
45b. _____
45c. _____

Score _____
(50 possible)

NAME: _____ DATE: _____

CHAPTER 10

Soldering and Brazing Principles and Practices

Choose the letter of the correct answer. If there is no lettered answer, write the word(s) answer.

1. Cooper pipe and tubing are used for plumbing and ____ installations.

    1. _____

2. Soldering requires higher temperatures than brazing. (True or False)

    2. _____

3. The flow of a liquid drawn into a small space between wet surfaces is called ____ action.

    3. _____

4. Capillary action is not a factor in the distribution of the brazing filler metal during braze welding. (True or False)

    4. _____

5. Selection of the proper material for soldering depends upon all but one of the following. Which one is incorrect?
    a. the metals to be joined
    b. the expected service
    c. the desired color of the weld
    d. the operational temperature of the weld

    5. _____

6. The ____ strength of a soldered joint refers to its ability to withstand long-term stress.

    6. _____

7. Match the types of solders with the most appropriate descriptive phrase.
    1. Tin lead   a. Most commonly used.
    2. Tin antimony   b. Non-toxic for use in food handling equipment.
    3. Tin zinc   c. Makes wide-clearance aluminum joints.
    4. Cadmium silver   d. Suitable for butt joints in copper tube.
    5. Cadmium zinc   e. In the proper ratio, melts and solidifies at the same temperature.

    7a. _____
    7b. _____
    7c. _____
    7d. _____
    7e. _____

8. Soldering fluxes do all but one of the following. Which one is incorrect?
   a. protect the base metal from oxidation
   b. improve filler metal flow
   c. increase wetting action
   d. clean the base metal
   e. float out oxides in the joint

8. _____

9. Fluxes may be classified in three groups. Which one of the following is incorrect?
   a. inorganic   b. organic   c. mineral   d. rosin

9. _____

10. Match the types of fluxes to the phrases which best describe them.
    1. Highly corrosive   a. Very active but highly volatile.
    2. Intermediate   b. May be localized at a joint to protect the rest of the work.
    3. Noncorrosive   c. Deposit chemically active residue on completed joint.
    4. Paste   d. Favored by electrical industry.

10a. _____
10b. _____
10c. _____
10d. _____

11. A clearance of ____ inch is recommended for capillary attraction to function well.

11. _____

12. The tensile strength of a joint increase as the joint clearance increases. (True or False)

12. _____

13. The heat required for soldering may be supplied by all but one of the following. Which one is incorrect?
    a. an electric arc   b. a torch
    c. an oven   d. induction   e. resistance

13. _____

14. The lowest flame temperature is produced by ____.
    a. propane   b. manufactured gas
    c. butane   d. natural gas   e. acetylene

14. _____

15. A sooty flame may prevent the solder from flowing smoothly by depositing ____ on the base metal.

15. _____

16. Pipe to be soldered or brazed is usually mechanically cleaned. (True or False)

16. _____

17. A tube may be too clean for effective soldering. (True or False)

17. _____

18. Production operations usually require ____ cleaning.

18. _____

NAME: _____ DATE: _____

19. The flux preferred for soldering copper tubing is ____. 19. _____
    a. highly corrosive   b. intermediate
    c. noncorrosive   d. paste

20. When soldering copper tubing ____. 20. _____
    a. prepare and flux joints at least a day before soldering
    b. heat as large an area as possible
    c. apply flame directly to solder until it is melted
    d. immediately pour water over the finished joint to wash away corrosive flux

21. The entire ____ are of a well-soldered joint will be covered with solder. 21. _____

22. Poorly soldered areas may be caused by all but one of the following. Which one is incorrect? 22. _____
    a. removing the torch flame from the solder pool
    b. poor cleaning
    c. improper fluxing
    d. loose fittings

23. Spelter is used for brazing copper tubing. (True or False) 23. _____

24. Compared to soldering, brazing does three of the following. Which is not correct? 24. _____
    a. uses higher temperatures
    b. requires special fluxes
    c. has less of an annealing effect
    d. forms a stronger joint

25. The minimum temperature at which brazing will take place is called the ____ temperature. 25. _____

26. The American Pipe Society lists classifications of brazing filler metals. (True or False) 26. _____

27. For brazed joints three of the following are true. Which one is incorrect? 27. _____
    a. are stronger than threaded ones
    b. are 90% as strong as the actual fittings
    c. are not loosened by vibration
    d. do not leak

28. Brazing filler metals that require the use of a flux include three of the following. Which one is incorrect? 28. _____
    a. aluminum silicon   b. copper phosphorous
    c. copper   d. nickel

CHAPTER 10: Soldering and Brazing Principles and Practices | 45

29. For joining parts in electron tube assemblies, ____ filler metals are used.

30. Except for some metals with low melting points, virtually all metals can be brazed with ____ alloys.

31. At ____ degrees F., flux loses its protective qualities.

32. Hard water cannot be effectively used in a flux. (True or False)

33. Brazing satisfactorily joins only similar metals. (True or False)

34. Lap joints are stronger than butt joints. (True or False)

35. Filler metals used for brazing are better electrical conductors than the copper tubing they join. (True or False)

36. Use ____ joints to fabricate pressure-tight assemblies.

37. The lowest flame temperatures are achieved with ____.
    a. air-gas torches
    b. butane torches
    c. oxhydrogen torches
    d. oxyacetylene torches

38. The widest range of heat control is provided by ____ torches.

39. Only brazing filler metals, which may be used with ____, are suitable for torch heating.

40. It is difficult to braze cast iron. (True or False)

41. When fluxing copper tubing in preparation for brazing, do all the following except ____.
    a. apply flux only in brazing area
    b. apply flux immediately after cleaning
    c. coat the tube inside and out
    d. always apply with a clean brush

42. Cooper tubing has reached the brazing temperature when the flux starts to bubble. (True or False)

43. Wrought copper fittings may be cooled quickly after brazing. (True or False)

29. _____

30. _____

31. _____

32. _____

33. _____

34. _____

35. _____

36. _____

37. _____

38. _____

39. _____

40. _____

41. _____

42. _____

43. _____

STUDENT WORKBOOK

NAME: _____   DATE: _____

44. All brazing fluxes must be removed after the brazing alloy has set. (True or False)

45. A pipe that is too large to heat adequately may be divided into ____ for brazing.

46. ____ destroys the effectiveness of the flux.

44. _____

45. _____

46. _____

Score _____
(53 possible)

NAME: _____ DATE: _____

CHAPTER 11

Shielded Metal-Arc Welding Principles

Choose the letter of the correct answer. If there is no lettered answer, write the word(s) answer.

1. In shielded metal-arc welding, the molten weld pool is protected by a gaseous shield and a covering of ____.

   1. _____

2. In shielded metal-are welding, the heat is generated by an electric arc established between a ____ and the work.

   2. _____

3. The nature of the metal electrode alone determines the chemical and physical characteristics of a shielded metal-arc weld. (True or False)

   3. _____

4. The shielded arc process is not suitable for welding metals with low melting points, such as three of the following. Which one is incorrect?
   a. zinc   b. lead   c. stainless steel   d. tin

   4. _____

5. The most dominant markets for SMAW usage is ____ and Maintenance and Repair.

   5. _____

6. The temperature of the electric arc has been measured as high as ____ degrees F.

   6. _____

7. Identify the elements of the welding circuit shown in Fig. 11-1.

   7a. _____
   7b. _____
   7c. _____
   7d. _____
   7e. _____
   7f. _____

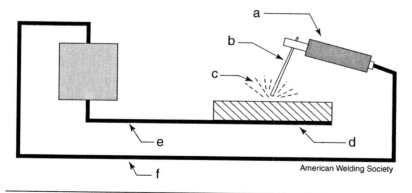

FIG. 11-1.

8. The output slope of welding machines may be classed as either constant voltage or constant ____.

8. _____

9. Which one of the following is not a type of welding machine?
   a. a.c. transformers
   b. volt-amperage rectifiers
   c. transformer-rectifiers
   d. motor generators
   e. inverter

9. _____

10. Inverter welding machines have no disadvantage. (True or False)

10. _____

11. When the arc is struck and the welding load is on the machine, open circuit voltage drops to ____ voltage.

11. _____

12. Lengthening the arc ____.
    a. increases both arc voltage and current
    b. decreases both arc voltage and current
    c. increases arc voltage and decreases current
    d. decreases arc voltage and increases current

12. _____

13. Motor-generator welding machines usually produce ____ current.

13. _____

14. Machine sizes are determined on the basis of ____.

14. _____

15. The actual output of a welding machine is always higher than the rated output. (True or False)

15. _____

16. A ____ switch is used to change the direction of the flow of welding current.

16. _____

17. Volt-ampere meters may indicate polarity. (True or False)

17. _____

18. Voltage measures the ____ of an electric current.

18. _____

19. Amperage measures the ____ of an electric current.

19. _____

20. The ability to strike and maintain a constant arc is determined by a welding machine's ____.

20. _____

21. ____ control makes it possible to adjust the welding machine without leaving the workstation.

21. _____

22. On a d.c. transformer-rectifier, the unit that changes the a.c. current from the power source to d.c. welding current is the ____.

22. _____

50 | STUDENT WORKBOOK

NAME: _____   DATE: _____

23. There are transformer-rectifier welding machines that produce both a.c. welding current and d.c. welding current. (True or False)

23. _____

24. A.c. welding machines have gained in popularity because of three of the following advantages. Which one is incorrect?
    a. improved machine performance
    b. noiseless operation
    c. adaptability to the use of large electrodes
    d. decreased power consumption
    e. adaptability to both straight and reverse polarity

24. _____

25. The two basic types of a.c. welding machines are the ____ type and the motor-generator type.

25. _____

26. The most important consideration in determining which type of welding machine to buy is the initial purchase cost. (True or False)

26. _____

27. To complete the circuit between a welding machine and the work, an ____ cable and a ground cable are required.

27. _____

28. Welding cable should be stiff so that it will wear well. (True or False)

28. _____

29. To determine the size of cable to choose for welding, consider machine ____ and distance from the work.

29. _____

30. Resistance to the flow of the welding current increases as the diameter of the cable increases. (True or False)

30. _____

31. The electrode holder used for shielded metal-arc welding may also be used for carbon-arc welding. (True or False)

31. _____

32. One of the disadvantages of atomic-hydrogen welding is that the welds are covered with scale. (True or False)

32. _____

33. When a number of welders are working close to each other, it is necessary to wear ____ behind the welding hood.

33. _____

34. Safety glasses can absorb more than 99.9 percent of harmful ultraviolet rays. (True or False)

34. _____

35. Chrome ____ helmets are used in cramped areas where there is not enough room for a standard hard helmet.

35. _____

CHAPTER 11: Shielded Metal-Arc Welding Principles | 51

36. Electronic filter welding helmets are not useful when working in close quarters. (True or False)

36. _____

37. Match the pieces of welding equipment to their best descriptions.
    1. Electrode holder
    2. Cable
    3. Electrode core
    4. Rotary clamp
    5. Lug
    6. Rectifier
    7. Transformer-rectifier machine

    a. Connects cable to welding machine.
    b. Supplies the weld metal.
    c. Can produce direct or alternating current.
    d. Turns with work without twisting cable.
    e. Changes alternating current to direct current.
    f. Carries electric current from the welding cable to the electrode.
    g. Completes electrical circuit between the welding machine and the work.

37a. _____
37b. _____
37c. _____
37d. _____
37e. _____
37f. _____
37g. _____

38. Match the welding terms to their best descriptions.
    1. Penetration
    2. Voltage
    3. Amperage
    4. Arc
    5. Output slope
    6. Ground

    a. Volt-ampere characteristic.
    b. Generates heat for welding.
    c. Depth of the weld into the plate.
    d. Measure of the amount of electrical current.
    e. Completes electrical circuit to protect welder from accidental shock.
    f. Force with which electrical current flows.

38a. _____
38b. _____
38c. _____
38d. _____
38e. _____
38f. _____

39. The clothing of choice for welders should be 100 percent cotton or wool. (True or False)

39. _____

Score _____
(55 possible)

NAME: _____ DATE: _____

CHAPTER 12

Shielded Metal-Arc Welding Electrodes

Choose the letter of the correct answer. If there is no lettered answer, write the word(s) answer.

1. Shielded metal-arc welding may be correctly referred to as all but one of the following. Which one is incorrect?
   a. arc welding   b. SMAW
   c. laser welding   d. stick electrode welding

   1. _____

2. ____ is a composite filler metal electrode consisting of a core of a bare electrode or metal cored electrode to which a covering sufficient to provide a slag layer on the weld metal has been applied.

   2. _____

3. The type of electrode covering influences the degree of penetration and the depth of the crater. (True or False)

   3. _____

4. The addition of large amounts of iron power to the coating of an electrode increases the speed of welding and improves the weld appearance. (True or False)

   4. _____

5. As a rule, when welding with a SMAW electrode, the maximum arc length is never greater than the diameter of the bare end of the electrode. (True or False)

   5. _____

6. Molten slag not only protects the weld metal, but also does three of the following. Which one is incorrect?
   a. removes oxides and impurities
   b. speeds up the cooling rate of the molten metal
   c. slows down the cooling rate of the solidified weld metal
   d. controls the shape and appearance of the weld

   6. _____

7. The electrode covering may be used to introduce additional elements into the weld metal. (True or False)

   7. _____

8. The welding of high carbon and alloy steels, high sulfur steels, and phosphorous-bearing steels has been improved by the introduction of low ____ electrodes.

   8. _____

9. The covering on a welding electrode makes it possible for four of the following. Which one is incorrect?
   a. to start the arc more easily
   b. to vary arc length
   c. to use either polarity
   d. work with higher currents
   e. to control the arc more effectively

9. _____

10. The core-wire composition of mild steel electrodes is changed depending upon the kind of covering used on the electrode. (True or False)

10. _____

11. Four of the following electrode coverings are correctly paired with the purposes they serve. Which one is incorrect?
    a. ferro-alloys: deoxidizers
    b. alkaline earth metals: provide reducing gases
    c. silica, clay: fluxes
    d. iron ore, mica: stagging ingredients
    e. potassium silicates: binders

11. _____

12. The composition of the covering determines the best polarity to use for d.c. applications. (True or False)

12. _____

13. The first two numbers in the AWS electrode classification indicate ____.
    a. polarity   b. tensile strength   c. welding current
    d. position to welding   e. covering

13. _____

14. The next to the last number in the AWS electrode classification indicates ____.
    a. polarity   b. covering   c. position of welding
    d. tensile strength   e. welding current

14. _____

15. The last number in the AWS electrode classification indicates the type of ____ to be used and the covering on the electrode.

15. _____

16. Selecting the proper size of electrode to use on a job depends upon all but which one of the following?
    a. joint design   b. material thickness
    c. size of welding machine   d. thickness of weld layers
    e. welding position required

16. _____

NAME: _____ DATE: _____

17. According to textbook Tables 12-6 and 12-7, E6012   17. _____
    electrodes may be used for all but one of the following.
    Which one is incorrect?
    a. only for welding in the vertical and overhead
       positions
    b. with direct current
    c. with alternating current
    d. for joints with poor fit-up

18. Certification in the F4 group designation does not   18. _____
    qualify in any other group. (True or False)

19. Match the electrode operating characteristics to their best
    descriptions.
    1. Fast fill   a. Deposits weld metal which   19a. _____
                            solidifies rapidly.
    2. Fast follow   b. Designed for fast downhand   19b. _____
                            welding and easy slag removal.
    3. Fast freeze   c. Produces ductile welds free   19c. _____
                            from underbead and
                            microcracking.
    4. Low hydrogen   d. Contributes additional metal to 19d. _____
                            weld deposit.
    5. Iron powder   e. Designed for vertical position, 19e. _____
                            travel-down welding and
                            reduced burn-through.

20. Match the types of electrodes to their operating
    characteristics.
    1. EXX12   a. Fast fill.   20a. _____
    2. EXX24   b. Low hydrogen.   20b. _____
    3. EXX10   c. Fast follow.   20c. _____
    4. EXX18   d. Fast freeze.   20d. _____

21. According to Table 12-10 in the text, of the following   21. _____
    metals the one with the lowest melting point is ____.
    a. zinc   b. steel   c. copper   d. silver   e. lead

22. According to textbook Tables 12-11 through 12-13, for   22. _____
    cast iron three of the following are true. Which one is
    incorrect?
    a. forms tough but breakable slag
    b. melts at a moderate speed
    c. can be made to form continuous chips
    d. makes a molten puddle that does not spark under
       blowpipe flame

23. In general never use an electrode having a diameter ____ than the thickness of the material being welded.

23. _____

24. For production work, the largest electrode size that can be handled should be used. (True or False)

24. _____

25. For overhead, vertical, or horizontal welding, EXX20 or EXX24 electrodes may be used. (True or False)

25. _____

26. For easier welding in the horizontal and flat positions, you can use electrodes classified in all but one of the following.
    a. EXX12   b. EXX20   c. EXX27   d. EXX28   e. EXX29

26. _____

27. To increase welding speed, use electrodes with ____ diameters.

27. _____

28. Match the electrode classifications to their best descriptions.
    1. E6010   a. Covering contains easily ionized materials.
    2. E6011   b. Best adapted for vertical and overhead welding.
    3. E6012   c. Often called the poor fitup electrode.
    4. E6013   d. Mineral-covered electrode does not need gaseous atmosphere protection.
    5. E6020   e. Requires alternating current to retain its best characteristics.

28a. _____
28b. _____
28c. _____
28d. _____
28e. _____

29. Tensile strength exceeding 100,000 psi may be obtained by welding with an electrode covered with alloy steel. (True or False)

29. _____

30. Stainless steel electrodes may contain both chromium and ____.

30. _____

31. For stainless steel electrodes four of the following are true. Which one is incorrect?
    a. resist corrosion   b. withstand high temperatures
    c. do not oxidize readily d. are easily cold worked
    e. resist scaling

31. _____

32. Match the stainless steel classifications to their best descriptions.
    1. Martensitic   a. Requires no heat treatment during welding.
    2. Ferritic   b. Magnetic, normally soft and ductile.
    3. Austenitic   c. Air hardening, normally hard and brittle.

32a. _____
32b. _____
32c. _____

NAME: _____ DATE: _____

33. SMAW electrodes are available for all PH stainless steel types. (True or False)

33. _____

34. ____ is the deposition of an alloy material on a metal part to form a protective surface.

34. _____

35. The effect described in question 34 can also be achieved by metal spraying. (True or False)

35. _____

36. E1100 and E4043 electrodes are the most commonly used ____ covered electrodes.

36. _____

37. Hot-shortness is a problem in the welding of aluminum. (True or False)

37. _____

38. The presence of ____ in the electrode covering is a major cause of a porous weld, underbead cracking, and rough appearance.

38. _____

39. All mineral-covered electrodes are thirsty. The minute they are unpacked, they start absorbing moisture—too much moisture for a sound weld. (True or False)

39. _____

40. Although it is recommended, over storage is not essential for preserving the quality of hard-surfacing electrodes. (True or False)

40. _____

Score _____
(53 possible)

NAME: _____ DATE: _____

CHAPTER 13

Shielded Metal-Arc Welding Practice: Jobs 13-J1-J28 (Plate)

Choose the letter of the correct answer. If there is no lettered answer, write the word(s) answer.

1. SMAW accounts for less than 50% of the total amount of welding being done. (True or False)

    1. _____

2. For stick electrode welding all but one of the following are true. Which one is incorrect?
   a. versatile
   b. dependable
   c. expensive but fast
   d. applicable to all welding positions

    2. _____

3. When welding, be as comfortable as possible in order to avoid ____.

    3. _____

4. Electrode positive will produce the deepest penetrating weld. (True or False)

    4. _____

5. In a d.c. circuit, the hottest side is the one with positive polarity. (True or False)

    5. _____

6. When welding is performed in electrode negative, the electrode lead becomes the negative side of the circuit and the ground lead becomes the positive. (True or False)

    6. _____

7. A fairly short arc is preferred for metal-arc welding in all but one of the following. Which one is incorrect?
   a. concentrates heat on the work
   b. reduces arc blow
   c. protects the weld pool from contamination
   d. blows out the flame periodically to rest the welder

    7. _____

8. In comparison to other welding positions, a slightly longer arc may be maintained for welding that is ____.
   a. vertical   b. flat   c. horizontal   d. overhead

    8. _____

9. As arc length increases arc voltage ____.   9. _____

10. The type of coating on the electrode may affect the arc   10. _____
    length. (True or False)

11. The arc is too long in all but one of the following. Which   11. _____
    one is incorrect?
    a. the metal flutters
    b. slight explosions occur regularly
    c. there is a sharp crackling sound
    d. the metal is easily seen

12. For arc blow all but one of the following are true. Which   12. _____
    one is incorrect?
    a. is caused by magnetic forces
    b. is more of a problem with shielded electrodes
    c. may cause porosity
    d. may be controlled by changing the electrode's position

13. The ____ equivalent of the fractional electrode sizes   13. _____
    gives some idea of the heat setting.

14. A 1/4-inch electrode requires higher current than a   14. _____
    3/8-inch electrode. (True or False)

15. Use more current for all but one of the following. Which   15. _____
    one is incorrect?
    a. a good head conductor
    b. electrode negative electrodes
    c. vertical-position welding
    d. larger pieces of material.

16. Current requirements vary with welding speed.   16. _____
    (True or False)

17. Holding the electrode at a proper ____ to the work   17. _____
    prevents undercut and slag inclusions.

18. Practice with bare electrodes allows a student to do all   18. _____
    but which one of the following?
    a. observe what takes place in the arc
    b. avoid distraction caused by arc blow
    c. observe fusion of weld metal and bare metal
    d. develop a steady hand for maintaining arc gap

NAME: _____ DATE: _____

19. Match the following welding problems to possible methods of correcting them.
    1. Spatter
    2. Undercut
    3. Poor fusion
    4. Porosity
    5. Arc blow

    a. Choose an electrode small enough to reach the bottom of narrow V's.
    b. Whitewash parts next to weld area.
    c. Use a.c. current if possible.
    d. Make a puddle to keep the weld molten longer.
    e. Avoid creating too large a weld puddle with an oversized electrode.

    19a. _____
    19b. _____
    19c. _____
    19d. _____
    19e. _____

20. To maintain the flow of electric current the electrode should be kept lightly touching the plate. (True or False)

    20. _____

21. Traveling too fast when making continuous stringer beads produces beads with all but one of the following. Which one is incorrect?
    a. are narrow
    b. show little reinforcement
    c. overlap
    d. have coarse ripples

    21. _____

22. At the completion of the weld bead fill craters by ____ the arc.

    22. _____

23. For deep groove joints, deposit beads wider than stringer beads by ____ the electrode from side to side.

    23. _____

24. Use a ____ weld to make a lap joint.

    24. _____

25. Match the electrode classifications to their characteristics.
    1. E6010   a. Flat position, DCSP, DCRP, or a.c.
    2. E6012   b. General purpose, reverse polarity.
    3. E6027   c. Low hydrogen, all positions.
    4. E7028   d. Iron powder, flat and horizontal positions.
    5. E7016   e. General purpose, straight polarity.

    25a. _____
    25b. _____
    25c. _____
    25d. _____
    25e. _____

26. The strength of a fillet weld depends upon its ____ thickness.

    26. _____

27. Welds made with bare electrodes are as strong and ductile as the base metal. (True or False)

    27. _____

CHAPTER 13: Shielded Metal-Arc Welding Practice: Jobs 13-J1-J28 (Plate)

28. Molten metal combines with gases in the air to form oxides and ____.

28. _____

29. Shielded-arc electrodes produce weld metal that is superior to the base metal. (True or False)

29. _____

30. Arc blow is a problem with covered electrodes because of the covering on the electrode. (True or False)

30. _____

31. Industry tends to use larger electrodes than it is recommended to begin practicing with. (True or False)

31. _____

32. E6013 electrodes are not suited to weld thin metals. (True or False)

32. _____

33. It is easier to weld in the vertical and overhead positions with a.c. electrodes. (True or False)

33. _____

34. Compared to direct welding current, all but one of the following are true of alternating current. Which one is incorrect?
    a. avoids arc blow
    b. uses larger electrodes
    c. strikes an arc more easily
    d. welds heavy steels faster

34. _____

35. Too short an arc gap causes ____.
    a. rough welds
    b. slag inclusions
    c. spatter
    d. oxidation

35. _____

36. The crater can be filled only with one technique. (True or False)

36. _____

37. Job 13-J6 calls for a ____-inch-high reinforcement bead.

37. _____

38. It is not good practice to weave beads wider than ____ times the diameter of the coated electrode.

38. _____

39. For E6010 electrodes all but one of the following are true. Which one is incorrect?
    a. produce high quality welds in all positions
    b. have a deep penetrating spray-type arc
    c. produce no spatter
    d. form a quick-setting weld pool

39. _____

40. E6011 electrodes produce weld metal with higher ductility and tensile strength than that produced by E6010 electrodes. (True or False)

40. _____

NAME: _____ DATE: _____

41. The arc gap for an E6010 electrode may be less than for an E6012 electrode. (True or False)

41. _____

42. Industrial weaved beading is usually done with ____ electrodes.

42. _____

43. Compared to DCEN E6010 and E6011 electrodes three of the following are true. Which one is incorrect?
    a. require higher bead reinforcement
    b. penetrate deeper
    c. form less slag
    d. solidify more quickly.

43. _____

44. Edge joints are used extensively for high-pressure vessel work. (True or False)

44. _____

45. For edge joints three of the following are true. Which one is incorrect?
    a. strong   b. economical
    c. easy to weld   d. air carbon-arc welded

45. _____

46. The edge joint is economical to set up. (True or False)

46. _____

47. For welding an edge joint an electrode should be held at a 90′ angle to the plate. (True or False)

47. _____

48. The lap joint is strongest when it is double-lap welded on both sides. (True or False)

48. _____

49. Lap joints require little preparation and fitup. (True or False)

49. _____

50. The top edge of the top plate of a lap joint has a tendency to burn away. (True or False)

50. _____

51. Travel that is too ____ will deposit excess weld metal on the plate.

51. _____

52. E6010 electrodes have a greater tendency to undercut than E6013 electrodes. (True or False)

52. _____

53. Lap joints are welded with ____ welds in the flat position.

53. _____

54. Stringer beads are used to weld horizontal ____ welds and fillet welds.

54. _____

55. For making horizontal stringer beads the plate is in the ____ position.

55. _____

56. In contrast to flat welding of stringer beads horizontal beading requires all but which one of the following?
    a. lower current
    b. longer arc
    c. electrode movement to avoid sagging
    d. bead overlap

56. _____

57. Welding downhill on a vertical surface is used for critical pipeline and sheet-metal work. (True or False)

57. _____

58. Vertical down welding requires a ____ arc gap.

58. _____

59. Downhill vertical welding requires a fast rate of travel. (True or False)

59. _____

60. Downhill vertical lap joint welds are used when maximum reinforcement and strength are required. (True or False)

60. _____

61. Undercutting is not a problem when downhill vertical welds are made with straight polarity electrodes. (True or False)

61. _____

62. For tank and structural work, it is often necessary to make lap welds in the ____ position.

62. _____

63. A T-joint must be welded with a ____ weld.

63. _____

64. For multipass fillet welds no more than ____-inch thickness of weld metal should be deposited on each layer.

64. _____

65. Most multipass welding is done with DCEP electrodes. (True or False)

65. _____

66. When making a weaved fillet weld with E6010 electrodes, hesitate longer at the sides of the weld than you would with electrode negative electrodes. (True or False)

66. _____

64 | STUDENT WORKBOOK

NAME: _____ DATE: _____

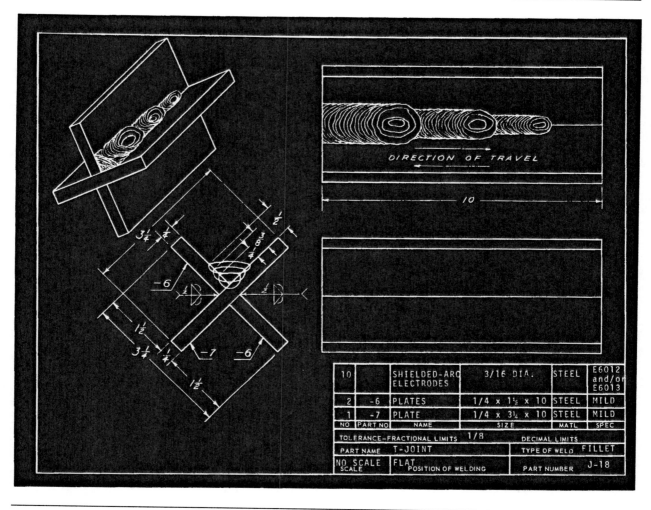

FIG. 13-1.

67. Refer to the print in Fig. 13-1 to match the measurements
   to the appropriate descriptions.
   1. Plate thickness   a. 1 1/2"
   2. Plate # 7 width   b. 1/4"   67a. _____
   3. Plate # 6 width   c. 1/2"   67b. _____
   4. Leg length of second fillet pass d. 3/8"   67c. _____
   5. Leg length of third fillet pass   e. 3/16" 67d. _____
   6. Size of electrode used   f. 3 1/4"   67e. _____
                                                  67f. _____

68. Interrupt multipass fillet welding to remove slag
    between each pass. (True or False)   68. _____

69. Electrode negative electrodes are generally used for
    vertical welding with travel up. (True or False) 69. _____

CHAPTER 13: Shielded Metal-Arc Welding Practice: Jobs 13-J1-J28 (Plate) | 65

70. When making a vertical weld, lengthen the arc gap on the upward stroke to allow deposited weld metal to solidify. (True or False)

70. _____

71. When welding in the vertical position calls for weaving a bead, travel is usually down. (True or False)

71. _____

72. If electrode negative electrodes are used for weave beading in the vertical position with upward travel, speed of travel should be ____ than with reverse polarity.

72. _____

73. Much of the welding done in the field is ____ fillet welding.

73. _____

74. The electrodes recommended for the job in Fig. 13-2 are ____.

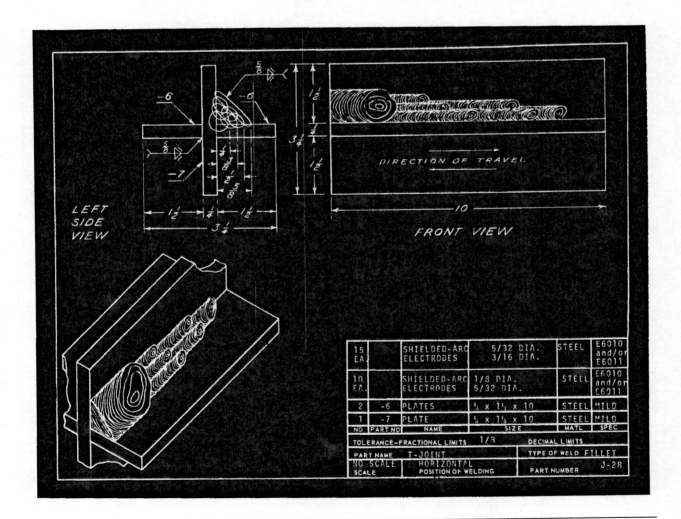

FIG. 13-2.

66 | STUDENT WORKBOOK

NAME: _____ DATE: _____

75. The electrode sizes recommended for the job in Fig. 13-2 include all but which one of the following.
   a. 1/4"   b. 1/8"   c. 3/16"   d. 5/32"

76. Refer to Fig. 13-2 to match the measurements to the appropriate descriptions.
   1. Limits of tolerance in fractions   a. 1/4"
   2. Print number   b. 1/8"
   3. Size of first fillet weld   c. 3/8"
   4. Size of passes 2 and 3   d. 5/8"
   5. Size of passes 4, 5, and 6   e. 1/2"
   6. Size of lacing bead   f. J-28

77. The print in Fig. 13-2 calls for a ____ welding position.

78. The part to be welded in Fig. 13–2 is a ____.

74. _____
75. _____

76a. _____
76b. _____
76c. _____
76d. _____
76e. _____
76f. _____

77. _____

78. _____

Score _____
(96 possible)

CHAPTER 13: Shielded Metal-Arc Welding Practice: Jobs 13-J1-J28 (Plate)

NAME: _____ DATE: _____

CHAPTER 14

Shielded Metal-Arc Welding Practice: Jobs 14-J26-J42 (Plate)

Choose the letter of the correct answer. If there is no lettered answer, write the word(s) answer.

1. For economic reasons and quality reasons, in the shop every effort is made to position the work so that welding can be in the flat or horizontal positions. (True or False)

    1. _____

2. For overhead welding all but one of the following is true. Which one is incorrect?
    a. is usually done with reverse polarity
    b. requires travel away from the operator
    c. is more tiring than other positions
    d. requires a longer arc gap with E6013 electrodes are used

    2. _____

3. If you are standing, drape the heavy welding cable over your shoulders to reduce the weight on the arm you are welding with. (True or False)

    3. _____

4. The whipping technique should be used with low hydrogen type electrodes (E7016-E7018). (True or False)

    4. _____

5. Backing material to a single-V butt joint creates the effect of welding a joint with a ____ ring.

    5. _____

6. Multipass welding of a single-V butt joint with a backing bar requires complete fusion with all but which one of the following?
    a. backing bar   b. bevel plates
    c. preceding passes   d. slag remnants

    6. _____

7. Refer to the ____ of the beveled plate to determine the width of the last pass.

    7. _____

8. The completed weld of a single-V butt joint should have a reinforcement of about ____ inch.

    8. _____

CHAPTER 14: Shielded Metal-Arc Welding Practice: Jobs 14-J26-J42 (Plate) | 69

9. The strength of a square butt joint welded from only one side depends upon all but one of the following. Which one is incorrect?
   a. penetration
   b. size of electrode used
   c. speed of welding
   d. amount of current

9. _____

10. For outside corner joints which one of the following is not true?
    a. are seldom used because of difficult preparation
    b. require welding conditions similar to those for V-groove butt joints
    c. require full penetration through the backside
    d. are more efficient with an inside corner fillet weld

10. _____

11. Look at Fig. 14-1 to match the following measurements to the appropriate descriptions.
    1. Surface plate thickness   a. 3/16
    2. Backup plate thickness   b. 10
    3. Joint length   c. 1 1/2
    4. Composite joints' width   d. 3/8
    5. Root, opening   e. 7
    6. Number of stringer beads required f. 8 1/4
    7. Width of surface plates   g. 5/32

11a. _____

11b. _____

11c. _____

11d. _____

11e. _____

11f. _____

11g. _____

NAME: _____   DATE: _____

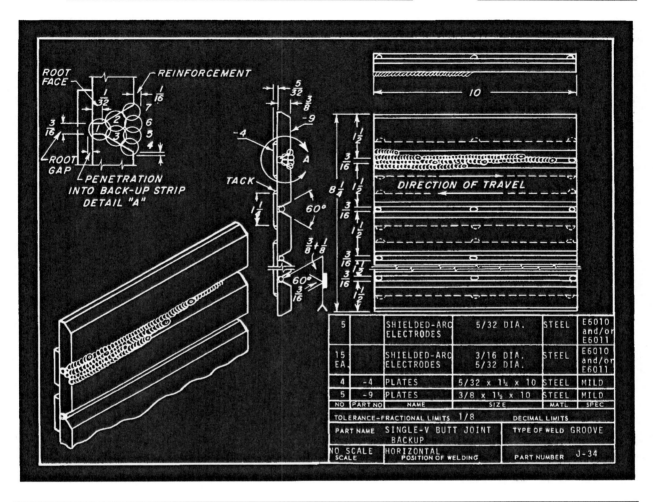

FIG. 14-1.

12. In welding the first pass in a flat position, single-V butt joint all but one of the following are correct. Which one is incorrect?
   a. hesitate at the sides of the weld
   b. keep the electrode within the space provided by the root gap
   c. deposit droplets of metal ahead of the weld
   d. maintain a small hole at the leading edge of the weld crater
   e. form a small bead on the reverse side of the groove

12. _____

13. The second pass of a single-V butt joint must achieve fusion with both the underneath pass and the ____ walls of the plate.

13. _____

14. All low hydrogen electrodes produce welds that are practically free of ____.

14. _____

15. For E7016 electrodes all but one of the following are true. Which one is incorrect?
    a. reduce underbead on low alloy steels
    b. are recommended for welds to be porcelain enameled
    c. produce welds suitable for nearly all code work
    d. use preheating to reduce stress requirements

15. _____

16. For F7016 electrodes all but one of the following are true. Which one is incorrect?
    a. use DCEP for their larger diameters
    b. are not suitable for vertical welding in their larger sizes
    c. produce rapidly freezing weld metal
    d. have a quiet arc with medium penetration and little spatter

16. _____

17. Match the measurements from Fig. 14-2 to the appropriate descriptions.
    1. First-pass electrode sizes
    2. Plate gap
    3. Back-side reinforcement on first pass
    4. Finish-pass reinforcement
    5. Finish-pass overlap

    a. 3/32"-1/8"
    b. 1/16"
    c. 1/8" or 5/32"
    d. 1/8"
    e. 3/32"

    17a. _____
    17b. _____
    17c. _____
    17d. _____
    17e. _____

NAME: _____   DATE: _____

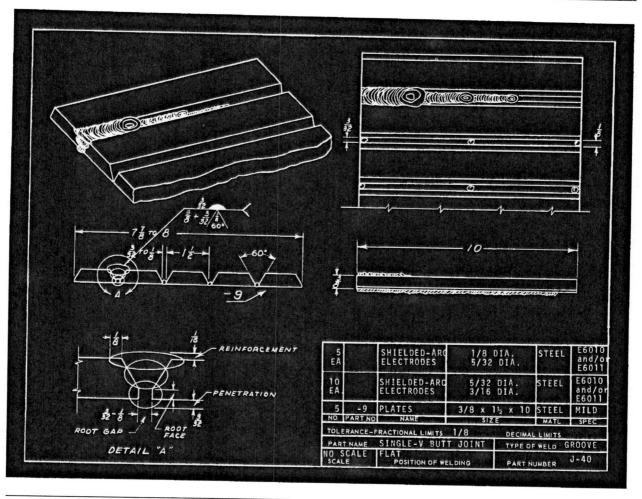

FIG. 14-2.

18. When making a vertical weld, travel up, ____ will make the weld pool spill off and run down.

18. _____

19. The whipping technique is used to help control the ____ penetration characteristics of cellulose E6010-E6011 electrodes.

19. _____

20. Multipass, vertical fillet welds are usually welded travel up with ____ electrodes.

20. _____

21. The iron powder in E7018 electrodes gives them all but one of the following advantages over the E7016 electrodes. Which one is incorrect?
    a. higher deposition rate,
    b. better restrike characteristics
    c. rougher bead
    d. more stable arc

22. Coatings containing iron powder are ____ than regular low hydrogen coatings.

23. Using an electrode with the coating in contact with the work is called ____ technique.

24. Downhill welding is ____ than the uphill method.

25. When downhill welding, keep the weld pool and slag from running ahead of the arc by using a shorter arc and ____ the rod angle and rate of travel.

21. _____

22. _____

23. _____

24. _____

25. _____

Score _____
(35 possible)

NAME: _____ DATE: _____

CHAPTER 15

Shielded Metal-Arc Welding Practice: Jobs 15-J43-J55 (Plate)

Choose the letter of the correct answer. If there is no lettered answer, write the word(s) answer.

1. It is necessary to read job prints and sketches to satisfy the skill demands of industrial shielded metal-arc welding jobs. (True or False)

2. Maximum strength in a single-V butt joint can be obtained only if there is complete ____ through the backside of the joint.

3. The whipping motion technique should never be used on low-hydrogen electrodes. (True or False)

4. When making a weaved bead in the vertical position, travel up, avoid excessive convexity in the center of the weld by moving rapidly across the ____ of the weld.

5. When depositing weld metal for a fillet weld in the overhead position, keep a short arc gap. (True or False)

6. A sound fillet weld in a T-joint should break evenly through the ____ when subjected to destructive testing.

7. A multipass fillet weld, in the overhead position is usually done with weaved beading. (True or False)

8. Each bead of a multipass fillet must be fused with the preceding ____ as well as the plate surfaces.

9. Each pass of a multipass weld requires a change in the ____ of the electrode.

10. A lap joint in the overhead position must be welded with stringer beads. (True or False)

11. Lap joints in the overhead position may be welded with DCEP or DCEN electrodes. (True or False)

1. _____
2. _____
3. _____
4. _____
5. _____
6. _____
7. _____
8. _____
9. _____
10. _____
11. _____

12. Although downhill welding of single-V butt joints is a rather difficult technique, once mastered, it is ____ than other methods.

    12. _____

13. Butt joints in pipe can be welded faster if a ____ is used.

    13. _____

14. For the first pass of a single-V butt joint, travel down, all but one of the following are true. Which one is incorrect?
    a. requires a close arc
    b. may be worked with a drag technique
    c. must be woven
    d. should produce no holes or burn-through

    14. _____

15. When welding a multipass single-V butt joint, travel down, which one of the following is correct?
    a. undercutting may be a problem
    b. hesitate at the sides of the weld to avoid undercut
    c. use a straight polarity electrode
    d. make the final bead somewhat convex

    15. _____

16. Holes or voids are acceptable in the second pass of a multipass single-V butt joint, travel down. (True or False)

    16. _____

17. When welding a T-joint in the overhead position with multipass fillets, break the arc and go back to the crater if it begins to solidify. (True or False)

    17. _____

18. A beveled-butt joint welded in the horizontal position is usually made with ____.
    a. laced beads
    b. stringer beads
    c. woven beads
    d. alternating stringer and weaved beads

    18. _____

19. Complete penetration is secured by using the correct combination of all but which one of the following?
    a. current
    b. arc gap
    c. electrode position
    d. travel speed

    19. _____

20. Welding a coupling to flat plate requires an electrode angle that is constantly ____ in relation to the welder.

    20. _____

21. When welding with a d.c. current, arc blow is not a possibility. (True or False)

    21. _____

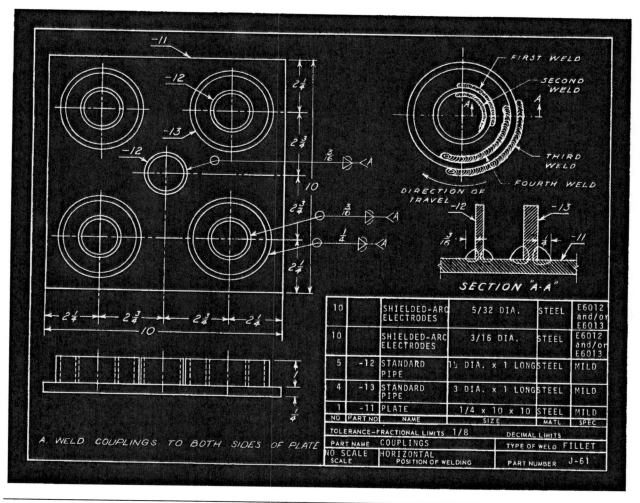

FIG. 15-1.

22. Match the measurements from Fig. 15-1 to the appropriate descriptions.

1. Smaller pipe diameter	a. 1/4"	22a. _____
2. Larger pipe diameter	b. 5 1/2"	22b. _____
3. Size of fillet on smaller pipe	c. 10"	22c. _____
4. Size of fillet on larger pipe	d. 1 1/4"	22d. _____
5. Distance from larger pipe center to nearest plate edge	e. 5"	22e. _____
6. Center-to-center distance of the larger couplings	f. 2 1/4"	22f. _____
7. Length and width of flat plate	g. 1/8"	22g. _____
8. Fractional tolerance limits	h. 3/16"	22h. _____
9. Distance from the center of the small, centered pipe to the plate edges	i. 3"	22i. _____

CHAPTER 15: Shielded Metal-Arc Welding Practice: Jobs 15-J43-J55 (Plate)

23. A beveled-butt joint in plate with a backup strip is often set up with a root gap of ____.
   a. 3/4"   b. 5/16"   c. 1/4"   d. 3/8"

23. _____

24. A single-V butt joint in the overhead position is usually welded with a electrode negative electrodes. (True or False)

24. _____

25. Weld test specimens are usually made of ____ steel.

25. _____

26. According to AWS, if a welder qualifies in one electrode group, he is also qualified in any lower group F-number. (True or False)

26. _____

27. A single-V butt joint is used to test ____ welds.

27. _____

28. In order to qualify, a welder must pass the tests in all positions. (True or False)

28. _____

29. Groove welds are subjected to face and root ____ tests.

29. _____

30. Test specimens are bent in a ____ to determine the soundness of the welds.

30. _____

31. In evaluating test results, no single indication shall exceed ____, measured in any direction on the surface.

31. _____

32. The pressure vessel test unit will begin to bulge at a p.s.i. pressure of ____.
   a. 300   b. 750   c. 1,200   d. 2,000   e. 2,800

32. _____

33. Air is used to provide the pressure for the vessel test unit. (True or False)

33. _____

34. A weld test specimen that shows any defects after bending is considered unsatisfactory. (True or False)

34. _____

35. The joints in Fig. 15-2 are indicated by circled numbers. Match numbers 1 through 4 to the types of joints, lettered a through d, which correctly identify them.
   a. Single-V butt   b. Lap
   c. Tee   d. Outside corner

35.1. _____
35.2. _____
35.3. _____
35.4. _____

78 | STUDENT WORKBOOK

NAME: _____ DATE: _____

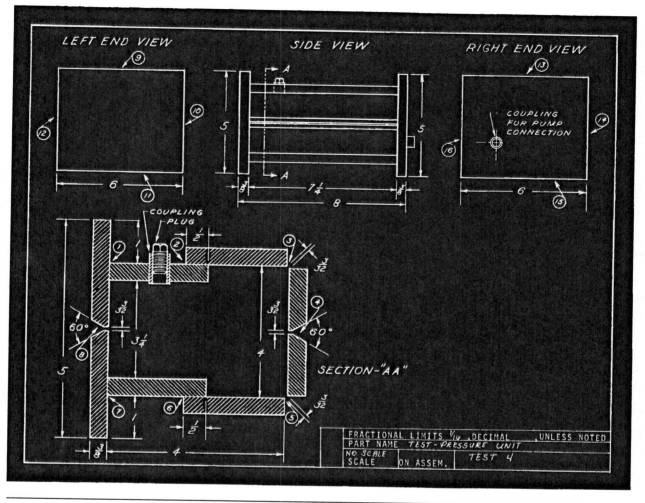

FIG. 15-2.

36. Match the following measurements from Fig. 15-2 to the
    appropriate dimensions.
    1. Length of end plates   a. 8"   36a. _____
    2. Width of end plates   b. 7 1/4"   36b. _____
    3. Distance between end plates c. 6"   36c. _____
    4. Overall unit length   d. 5"   36d. _____

                                               Score _____
                                                    (50 possible)

NAME: _____   DATE: _____

CHAPTER 16

Shielded Metal-Arc Welding Practice (Pipe)

Choose the letter of the correct answer. If there is no lettered answer, write the word(s) answer.

1. Pipe diameters over 20 inches cannot be welded satisfactorily. (True or False)

2. All but which one of the following ferrous metals are used to make pipe?
   a. nickel
   b. wrought iron
   c. stainless steel
   d. carbon steel

3. Nonferrous metals used in pipe construction include all but which one of the following?
   a. titanium
   b. copper
   c. asbestos
   d. aluminum

4. Power plants rely heavily on piping systems, pressure vessels and various other welded structures. (True or False)

5. The most common use for welded piping in buildings is hot water systems for ____.

6. Pipelines must be able to service low and high pressures and ____.

7. There are more miles of pipeline in the U.S. than there are miles of railroad track. (True or False)

8. At least ____% of the nation's overland cargo flows through pipelines.

9. Welded pipeline forms an underground network for transporting only petroleum products. (True or False)

10. Product pipelines transport more than ____ refined petroleum products.

1. _____

2. _____

3. _____

4. _____

5. _____

6. _____

7. _____

8. _____

9. _____

10. _____

11. It is possible to move a product through a pipeline with an average speed of three-eighths mile per hour. (True or False)

11. _____

12. Sections of pipe may be as long as ____ feet.

12. _____

13. Round rubber balls or similar separators used to keep different products flowing through pipelines apart are called ____.

13. _____

14. At the present time, practically all pipe 2 inches in diameter and larger are joined by welding. (True or False)

14. _____

15. All but one of the following are advantages of welded pipe installations. Which one is incorrect?
    a. permanence
    b. low maintenance costs
    c. heavier, more durable fittings
    d. improved flow characteristics

15. _____

16. If all the joints on a piping system are welded, the cost of materials alone may be reduced by ____%.

16. _____

17. Welded pipe fittings allow the use of ____ pipe walls with no loss of efficiency.

17. _____

18. The life expectancy of a welded pipe system is ____ times that of a threaded system.

18. _____

19. Thread cutting reduces pipe strength. (True or False)

19. _____

20. Inside diameters of welded fittings and pipe meet exactly to improve flow. (True or False)

20. _____

21. When compared with threaded systems, welded piping includes all but which one of the following?
    a. may be shop-fabricated
    b. requires slightly more equipment
    c. requires less space
    d. is more efficient

21. _____

22. The material most widely used in piping systems is ____.
    a. aluminum   b. carbon steel
    c. polyvinyl chloride   d. wrought iron

22. _____

82 | STUDENT WORKBOOK

NAME: _____ DATE: _____

23. High pressure transmission liens are made of low ____ steel pipe.

23. _____

24. Wrought iron pipe is preferred for all but which one of the following?
    a. underground systems
    b. condensate piping
    c. high-pressure lines
    d. radiant heating operations

24. _____

25. For extremely low or high temperatures, use ____ pipe.

25. _____

26. A special type of arc welding machine is required to weld pipe. (True or False)

26. _____

27. For general pipe welding, ____ electrodes are most commonly used.

27. _____

28. A type of electrode especially designed for vertical down welding belongs to the classification ____.
    a. E6010   b. E6011   c. E7010   d. E11018G

28. _____

29. The highest tensile strength is achieved in welds made with electrodes belonging to the classification ____.
    a. E6010   b. E6011   c. E7010   d. E11018G

29. _____

30. Pipe welds may achieve strengths as high as ____ psi.

30. _____

31. Which one of the following is not one of the joints most commonly used in welded piping systems?
    a circumferential butt   b. lap
    c. socket   d. lateral

31. _____

32. The types of welds most commonly used in welded piping systems are grooves and ____.

32. _____

33. Inner liners used as supports for butt joints may be called chill or ____ rings.

33. _____

34. All but which one of the following are true of pipe butt joints?
    a. are easily prepared
    b. may be welded in all positions
    c. are faultless except for some cracking
    d. allow good stress distribution

34. _____

CHAPTER 16: Shielded Metal-Arc Welding Practice (Pipe) | 83

35. Butt welds are usually applied pipe to all but which one of the following?
   a. socket   b. flanges   c. valves   d. pipe

35. _____

36. Use of the butt joint for pipe is limited to thick pipe. (True or False)

36. _____

37. The accepted bevels for welded industrial pipe range from 30 to ____ degrees.

37. _____

38. For socket joints all but one of the following are true. Which one is incorrect?
   a. are fillet welded
   b. are restricted to use in small pipe
   c. are recommended for radioactive systems
   d. are not suitable for exposure to corrosive materials

38. _____

39. For factory-manufactured welding pipe fittings all but one of the following are true. Which one is incorrect?
   a. equal the strength of the pipe being welded
   b. are welded in the horizontal position
   c. require only butt welds
   d. require less space than fabricated ones

39. _____

40. Chill rings assist the welder in securing complete penetration and fusion without ____.

40. _____

41. Backing rings prevent the formation of icicles or ____ inside the joint.

41. _____

42. Backing rings that expand or contract to fit the inside diameter of a pipe are called ____.

42. _____

43. Backing rings may have a series of small nubs around the ring to aid in spacing the pipe joint. (True or False)

43. _____

44. A consumable ____ ring improves weld quality, resists shrinkage and hot shortness, and eliminates weld-root notches.

44. _____

45. Fabricated weld fittings are formed from straight lengths of pipe by hand or machine ____ cutting.

45. _____

46. Piping codes have been established to insure uniform practices in design, installation, and ____.

46. _____

47. Some insurance companies, manufacturers, and the ____ set up their own welding fabrication codes.

47. _____

NAME: _____ DATE: _____

48. Match the welding societies to the types of welding for which they provide codes.
   1. American Society of Mechanical Engineers
   2. American Petroleum Institute
   3. American Standards Association
   4. American Water Works Association
   5. Heating, Piping, Air Conditioning Contractors
   6. Pipe Fabrication Institute
   7. American Welding Society

   a. Arc welding of piping for compressing and pumping fuel gases.
   b. Pioneer of training and testing practices.
   c. Piping connected to boilers.
   d. Joint preparation, preheating, and welding.
   e. Piping for water purification plants.
   f. Piping for heating.
   g. Pressure piping for the chemical industry and nuclear energy.

   48a. _____
   48b. _____
   48c. _____
   48d. _____
   48e. _____
   48f. _____
   48g. _____

49. Procedure qualification tests establish a procedure for producing welds of suitable mechanical properties and ____.

   49. _____

50. If any of the essential variables are changed beyond their limits the welder must requalify. (True or False)

   50. _____

51. The purpose of the welder qualification test is to determine the ability of the welder to make sound welds using a previously qualified welding procedure. (True or False)

   51. _____

52. Once a welder is qualified for one of the codes, he can do all types of work without taking any more tests. (True or False)

   52. _____

53. Test coupons that have been subjected to test procedure must not show imperfections of any kind. (True or False)

   53. _____

54. Changing the direction of travel for a procedure for which a welder has already qualified requires retesting. (True or False)

   54. _____

55. The highest paid welders are code ____ welders.

   55. _____

56. For a test weld, the least important of the following qualities is ____.
    a. penetration   b. appearance
    c. fusion   d. bevel angle

56. _____

57. Non-destructive testing methods include all but which one of the following?
    a. radiographic   b. magnetic particle
    c. ultrasonic   d. etching

57. _____

58. Visual inspection is actually a form of ____ testing and is best applied before, during and after welding is complete.

58. _____

59. Radiographic inspection is not used very much because it is expensive and unreliable. (True or False)

59. _____

60. Ultrasonic inspection should be interpreted by a highly experienced inspector. (True or False)

60. _____

61. Liquid penetrant inspection is used extensively on nonmagnetic materials. (True or False)

61. _____

62. Magnetic particle inspection can be used on both magnetic and nonmagnetic materials. (True or False)

62. _____

63. Compressed air is used for hydrostatic testing. (True or False)

63. _____

64. Removing a cylindrical plug from a pipe weld to inspect the inside surface is called ____.

64. _____

65. The variable in any welding procedure is the ____.

65. _____

66. The incomplete filling of the weld root with Weld metal is called incomplete ____.

66. _____

67. Incomplete ____ is the lack of fusion between beads or between the weld metal and base metal.

67. _____

68. Internal ____ is a condition in which the center of the bead lacks buildup and is below the inside surface of the pipe wall.

68. _____

69. Poor beveling, preparation, and setup can be as responsible for failure as poor welding technique. (True or False)

69. _____

NAME: _____ DATE: _____

70. An insufficient angle of bevel will make it very difficult to secure the proper root penetration and ____.

70. _____

71. Tack welds are used to hold a joint in proper alignment and to minimize ____.

71. _____

72. The root opening for a pipe joint should be from 3/32 to ____ of an inch.

72. _____

73. In tacking up a pipe joint, the easiest way to insure accurate root opening is to use gas-welding ____ as a spacer.

73. _____

74. Tack welds should never be more than an ____ inch long.

74. _____

75. Most cross-country pipelines are welded with vertical-____ travel.

75. _____

76. Compared to vertical-down welding, vertical-up does all but which one of the following?
a. uses lower current
b. uses less electrode per joint
c. cleans up faster
d. melts out gas holes more effectively

76. _____

77. When welding in the rolled or horizontal fixed position, one layer of weld metal for each ____ inch of pipe wall thickness is recommended.

77. _____

78. A pipe joint welded in the rolled or horizontal fixed positions requires a ____ inch electrode for the final passes.

78. _____

79. When pipe in the fixed vertical position is welded on a horizontal plane, the weld metal is deposited in a series of ____ stringer beads.

79. _____

80. The work connection from the welding machine should be attached as far from the pipe joint as possible. (True or False)

80. _____

81. Match the following welding excesses to the defects they cause in a finished weld.
    1. Arc gap too long
    2. Arc gap too short
    3. Slag coverage poor
    4. Travel speed too fast
    5. Current too high

    a. Excessive spatter, wide bead.
    b. Porosity.
    c. Undercutting; high, narrow bead.
    d. Bead pile up.
    e. Excessive penetration inside pipe.

    81a. _____
    81b. _____
    81c. _____
    81d. _____
    81e. _____

82. The hole formation at the tip of the root pass of a pipe butt weld is referred to as the ____.

    82. _____

83. Each pass of a multipass weld on a pipe joint must fuse with the underneath pass and the side ____ of the pipe.

    83. _____

84. To avoid undercut in the weave pass on a horizontal butt joint in the fixed vertical position, pause at the ____ of the weave.

    84. _____

85. The fixed horizontal position for a butt weld on pipe is referred to as the ____ position.

    85. _____

86. The root pass for a butt weld in the fixed horizontal position, travel up, should be started at the bottom center of the pipe. (True or False)

    86. _____

87. When restarting a weld bead with a new electrode, it is important that the start be made at the end of the previous weld. (True or False)

    87. _____

88. External ____ along the edges of a weld bead is called wagon tracts.

    88. _____

89. Insufficient penetration of a pipe weld may be caused by having all but one of the following. Which one is incorrect?
    a. excessive root face
    b. the root opening too narrow
    c. the electrode diameter too large
    d. the welding current too low

    89. _____

90. Internal undercut of a pipe joint will occur in all but which one of the following?
    a. the root face is too small
    b. the root opening is too large
    c. the welding current is too high
    d. the angle of bevel is too wide

    90. _____

NAME: _____ DATE: _____

91. The second pass on a pipe weld is often referred to as the ____ pass.

91. _____

92. A ____ pass is used to build thin sections up to the height of the rest of the weld bead.

92. _____

93. Because downhill welded beads are thinner than uphill ones, additional ____ passes are required to complete a joint.

93. _____

94. The cover pass is also referred to as the ____ pass.

94. _____

95. It is common practice in pipe welding to hand-fabricate as many pipe fittings as possible in the field. (True or False)

95. _____

Score _____
(105 possible)

CHAPTER 16: Shielded Metal-Arc Welding Practice (Pipe) | 89

NAME: _____ DATE: _____

CHAPTER 17

Arc Cutting Principles and Arc Cutting Practice

Choose the letter of the correct answer. If there is no lettered answer, write the word(s) answer.

1. The primary advantage of arc cutting is that it can be used on all types of metal. (True or False)

   1. _____

2. Plasma arc cutting is similar in many respects to what common welding process?

   2. _____

3. The fourth state of matter is ___.

   3. _____

4. The process which uses compressed air instead of gravity and the force of the arc to remove molten metal from the work is ___ carbon arc cutting.

   4. _____

5. The air carbon-arc process may be used to prepare plates for welding. (True or False)

   5. _____

6. The air carbon-arc process can be used for fabrication, but it is seldom used for repair and maintenance. (True or False)

   6. _____

7. The depth and contour of the groove when using CAC-A is controlled by all but which one of the following?
   a. electrode angle   b. material type
   c. travel speed   d. current

   7. _____

8. Both goggles and ___ must be worn during the cutting process for protection from sparks, metal particles, and heat.

   8. _____

9. To cut straight lines accurately, use ___.
   a. straight gouging squares
   b. magnetic burning tips
   c. rivet-burning tips
   d. slicing-gauging tips
   e. magnetic burning squares

   9. _____

10. A portable ___ cutter may be used for trimming and beveling angles and channels.

    10. _____

11. Any alloy with a high hardenability will be in a ____ condition after using the CAC process to cut the metal.

11. _____

12. What is the recommended current range for 3/16″ diameter carbon arc cutting electrode?

12. _____

13. Cuts performed in the overhead position with the CAC process are a hazard to the operator because of the molten ____.

13. _____

14. Oxygen-arc cutting surfaces are rougher than those produced by other arc cutting processes. (True or False)

14. _____

15. Oxygen-arc cutting cannot be used to cut stainless steel and aluminum. (True or False)

15. _____

16. A special holder is required for oxygen-arc cutting. (True or False)

16. _____

17. Fuel gases are not used with arc cutting. (True or False)

17. _____

18. All but which one of the following is true of carbon arc cutting?
    a. prepares metals for postheat treatment after welding
    b. makes a smooth, accurate cut
    c. produces a very hot arc
    d. allows long service from soft or hard carbon electrodes

18. _____

19. Arc cutting is superior to oxy-fuel gas cutting because of all but which one of the following?
    a. work is smoother and of better quality
    b. all types of metals can be cut
    c. the heat can be generated solely by electricity
    d. both carbon and metal arcs can be used

19. _____

20. Surfaces cut by the SMAC process are less ragged than those produced by the carbon arc. (True or False)

20. _____

21. The MIG welding process can also be used as a cutting process. (True or False)

21. _____

22. The nature of the MIG cut is influenced by the wire-speed feed and the ____.

22. _____

23. As voltage increases, kerf width ____.

23. _____

NAME: _____ DATE: _____

24. For low-speed MIG cutting, use ____ current.   24. _____

25. Bare metal-arc cutting is well-suited to cutting thick   25. _____
    metals. (True or False)

26. In the TIG cutting process, an electric arc is struck   26. _____
    between a ____ electrode and the metal workpiece.

27. The amount of current used for TIG cutting is about the   27. _____
    same as that used for TIG welding. (True or False)

28. The TIG process introduces all the advantages of oxygen   28. _____
    cutting to ____ metals and stainless steel.

29. The cutting quality of the TIG process can be improved   29. _____
    by using a shielding gas of ____ mixed with hydrogen.

30. To be cut by the plasma arc process, a metal must be able   30. _____
    to conduct ____.

31. Regarding the gas used in the plasma arc cutting process   31. _____
    all but one of the following are true. Which one is
    incorrect?
    a. produces a flame for cutting the work
    b. shields the operation from the air
    c. maintains an electric arc
    d. oxidizes the material being cut

32. The high pressure gas flowing through the plasma torch   32. _____
    is heated to a temperature of
    a. 2,700° F.   b. 5,000° F.   c. 35,000° F.
    d. 120,000° F.   e. 150,000° F.

33. Stainless steels and other nonferrous metals require a   33. _____
    ____ gas for plasma arc cutting.

34. Clean cuts can be achieved with the PAC process on   34. _____
    most metals up to ____ inches thick.

35. When cutting carbon steel with the PAC process,   35. _____
    superior results are achieved when nitrogen and ____ are
    used.

36. Stainless steel and nonferrous alloys are generally cut   36. _____
    with mixtures of argon and ____ when using the PAC
    process.

CHAPTER 17: Arc Cutting Principles and Arc Cutting Practice | 93

37. Magnesium can be cut with the PAC process. (True or False)

37. _____

38. The ____ arc is an arc between the electrode and the torch tip in the PAC torch.

38. _____

39. A major concern of ____ with PAC is that it may interfere with telephones, computers or CNC machines.

39. _____

40. PAC creates a large heat-affected zone. (True or False)

40. _____

Score _____
(40 possible)

NAME: _____ DATE: _____

CHAPTER 18

Gas Tungsten-Arc and Plasma Arc Welding Principles

Choose the letter of the correct answer. If there is no lettered answer, write the word(s) answer.

1. The GTAW process was originally developed for welding corrosion-resistant and other difficult-to-weld materials such as magnesium, stainless steel, and aluminum. (True or False)

    1. _____

2. Flux cored and metal-cored wires are expansions of what basic welding process?

    2. _____

3. In the TIG welding process, the tungsten electrode is consumed. (True or False)

    3. _____

4. It is always necessary to use filler rod when TIG welding. (True or False)

    4. _____

5. The shielding gases most commonly used for TIG welding are argon and ____.

    5. _____

6. For the gas metal-arc process all but one of the following are true. Which one is incorrect?
    a. is a nonconsumable electrode process
    b. can operate with a mixture of shield gases
    c. employs one wire as both electrode and filler metal
    d. uses carbon dioxide extensively

    6. _____

7. Because its arc is easily started, CIG welding is widely used in industry. (True or False)

    7. _____

8. Gas-shield arc spot-welding can be used to weld both ferrous and nonferrous metals. (True or False)

    8. _____

9. Gas-shielded arc spot-welding has all of the following advantages over resistance spot-welding except one. Which one is incorrect?
   a. access to only the front side of the joint is sufficient
   b. portable equipment can also be used for MIG and TIG welding
   c. heavy plates can be welded
   d. spatter, smoke, and sparks are minimal
   e. the workpiece suffers little distortion

10. Gas-shielded arc welds are stronger, more ductile, and more resistant to corrosion than those made by the shielded metal-arc process. (True or False)

11. Combinations of dissimilar metals can be welded with the gas-shielded arc process. (True or False)

12. In gas-shielded arc welding, the air in the arc area is displaced by inert gas before the arc is struck. (True or False)

13. The three elements in the air that cause the most weld contamination are oxygen, hydrogen, and ____.

14. Helium and argon must never be mixed for use in TIG welding. (True or False)

15. Both helium and argon are all but which one of the following?
    a. chemically inert
    b. suited to welding thick metal plates
    c. not damaging to tungsten electrodes
    d. effective mixed together

16. Of the two gases in question 15, ____ allows faster welding and deeper penetration.

17. Hydrogen may be mixed with helium for TIG welding. (True or False)

18. Nitrogen may be used in TIG welding to weld deoxidized ____.

19. Nitrogen can be used as a backup shield for the TIG welding of stainless steel. (True or False)

20. Helium provides more weld coverage than argon when welding in the vertical and overhead positions. (True or False)

9. _____

10. _____

11. _____

12. _____

13. _____

14. _____

15. _____

16. _____

17. _____

18. _____

19. _____

20. _____

NAME: _____ DATE: _____

21. ____ indicate the rate of flow of inert gas to the torch.   21. _____

22. Flowmeters are calibrated to indicate flows in all but which one of the following?   22. _____
    a. cubic feet per second
    b. cubic feet per minute
    c. cubic feet per hour
    d. cubic feet per inch

23. Too much gas at the end of the tungsten electrode can cause weld ____.   23. _____

24. The machines used for TIG welding are classified as all but which one of the following?   24. _____
    a. constant current and voltage
    b. variable current and voltage
    c. constant current, variable voltage
    d. variable current, constant voltage

25. The type of operating current directly affects weld ____, contour, and metal transfer.   25. _____

26. With the GTAW process, ____ frequency allows for the arc to be started without touching the electrode to the work.   26. _____

27. TIG welding electrodes may be all but which one of the following?   27. _____
    a. tungsten   b. thoriated
    c. striped   d. varigated

28. Gas nozzles used for TIG welding are made of refractory ____, glass, or metal.   28. _____

29. The TIG torch may be equipped with a gas ____ to prevent turbulence of the gas stream.   29. _____

30. The ____ feeds both the current and the inert gas to the weld zone for GTAW.   30. _____

31. The melting point of ____ is higher than that of all other elements except carbon.   31. _____

32. Pure tungsten electrodes are generally used for TIG welding with a.c. current. (True or False)   32. _____

CHAPTER 18: Gas Tungsten-Arc and Plasma Arc Welding Principles | 97

33. For thoriated tungsten electrodes all but one of the following are true. Which one is incorrect?
    a. may be used with a.c. current
    b. are less expensive to use than pure tungsten
    c. do not contaminate the work
    d. facilitate touch starting

34. Tungsten electrodes have a small ____ band to indicate their alloy.

35. The current range for a 3/32" diameter pure tungsten when GTAW with ACHF current and argon shielding gas is ____.

36. Grinding of a tungsten electrode should be done on a fine-grit, hard abrasive ____.

37. The length of the tapered end of a correctly ground tungsten should be ____.
    a. 1/2 times the tungsten diameter
    b. 1 times the tungsten diameter
    c. 1-1/2 times the tungsten diameter
    d. 4 times the tungsten diameter

38. The welding end of the tungsten electrode can be pointed or ____.

39. Tap water is generally not recommended for use as a coolant for GTAW torches because of its inherent ____ content.

40. Coolant flowing through the GTAW torch should go directly from the coolant source to the ____.

41. For hot wire welding all but one of the following are true. Which one is incorrect?
    a. is a TIG process
    b. requires preheated filler wire
    c. can be manual or mechanized
    d. concentrates arc heat directly on the weld

42. Welds made with the hot wire process do not require post-weld cleaning. (True or False)

43. TIG hot wire welding is not recommended for aluminum or ____.

33. _____
34. _____
35. _____
36. _____
37. _____
38. _____
39. _____
40. _____
41. _____
42. _____
43. _____

NAME: _____ DATE: _____

44. Plasma welding is done with a.c. current. (True or False)   44. _____

45. The ____ effect is unique to plasma welding.   45. _____

46. Plasma arc surfacing is a metal spray process. (True or False)   46. _____

47. Plasma arc surfacing minimizes base-metal dilution. (True or False)   47. _____

48. Plasma arc surfacing works best on ____ production jobs.   48. _____

49. The depth of penetration with plasma arc surfacing is precisely controlled to within ____.   49. _____

50. Because plasma arc surfacing uses ____ and alloys, it is not limited by wire availability.   50. _____

Score _____
(50 possible)

CHAPTER 18: Gas Tungsten-Arc and Plasma Arc Welding Principles | 99

NAME: _____ DATE: _____

CHAPTER 19

Gas Tungsten-Arc Welding Practice (Plate)

Choose the letter of the correct answer. If there is no lettered answer, write the word(s) answer.

1. TIG and MIG welding are recommended for aluminum because their ____ gas protects the weld pool.

    1. _____

2. When TIG welding aluminum all but one of the following are true. Which one is incorrect?
    a. there is no glare
    b. there is no smoke
    c. the weld pool is visible
    d. filler metal may be added outside the weld pool

    2. _____

3. The TIG process is preferred for welding aluminum sections up to 1/2-inch thick. (True or False)

    3. _____

4. The current preferred for TIG welding of aluminum is ____.
    a. DCSP   b. a.c. with stabilization
    c. DCRP   d. d.c. with low frequency

    4. _____

5. In addition to any other contamination on an aluminum plate, any ____ film must also be removed.

    5. _____

6. Backup strips for aluminum joints may be made of all but which one of the following?
    a. steel   b. copper
    c. stainless steel   d. wrought iron

    6. _____

7. For TIG welding aluminum, all but one of the following are true. Which one is incorrect?
    a. preheating is not always required
    b. distortion is seldom a problem
    c. the weld pool solidifies rapidly
    d. all welding positions may be used

    7. _____

8. When welding aluminum with the TIG process, rounding the end of the electrode will help ____ the arc.

    8. _____

9. For TIG welding carbon and low alloy steels all but one of the following are true. Which one is incorrect?
   a. is easier than welding stainless steels
   b. is used extensively for heavy gauge steels
   c. requires less heat than aluminum from a given electrode size
   d. requires a lighter shade of welding lens than other steels

9. _____

10. In the TIG welding of carbon steels, best results are obtained with all but which one of the following?
    a. DCSP
    b. argon shielding gas
    c. a balled electrode end
    d. a thoriated electrode

10. _____

11. The highest quality TIG welds require the use of a ____ control.

11. _____

12. Stainless steel is widely used for welds in all but one of the following. Which one is incorrect?
    a. for use under high temperatures
    b. to withstand high pressures
    c. with minimal thermal expansion
    d. resistant to corrosion

12. _____

13. For austenitic stainless steels all but which one of the following are true?
    a. belong to the 300 series
    b. are hardened by cold working
    c. are highly weldable
    d. may be welded only with direct current

13. _____

14. For martensitic stainless steels all but which one of the following are true?
    a. are somewhat brittle
    b. are straight chromium
    c. are hardened by rapid cooling from high temperatures
    d. belong to both 400 and 500 series

14. _____

15. For ferritic stainless steels all but which one of the following are true?
    a. are welded with high heat
    b. have high chromium content
    c. belong to the 400 series
    d. can be TIG welded

15. _____

NAME: _____   DATE: _____

16. TIG welding of stainless steel eliminates all but which one of the following?
    a. flux   b. spatter   c. heavy slag   d. distortion

16. _____

17. When making a TIG weld in stainless steel, greater penetration and welding speed can be obtained with ____ current.

17. _____

18. Both argon and ____ may be used as the shielding gas for the TIG welding of stainless steel.

18. _____

19. A mixture of argon and ____ is the equivalent of helium and produces sound welds in austenitic stainless steels.

19. _____

20. Magnesium is approximately which one of the following?
    a. 1/3 the weight of aluminum
    b. 2/3 the weight of aluminum
    c. 1-1/2 times the weight of aluminum
    d. the same weight as aluminum

20. _____

21. For copper which one of the following is incorrect?
    a. is similar to aluminum for welding purposes
    b. is strongest when cold worked
    c. most commonly uses DCEN for welding
    d. requires higher welding current settings than other metals

21. _____

22. For TIG welding nickel which one of the following is incorrect?
    a. eliminates slag entrapment
    b. uses current values in excess of those for carbon steel
    c. attains lower heat for thin materials from ACHF
    d. generally uses DCEN

22. _____

23. TIG welding is especially suited to welding plate thicknesses up to ____ inch.

23. _____

24. The backup for TIG welding may be ____ painted on the weld underside.

24. _____

25. The inert gas recommended for backing for stainless steels is ____.

25. _____

26. A 3/16-inch tungsten electrode is the largest size available. (True or False)

26. _____

CHAPTER 19: Gas Tungsten-Arc Welding Practice (Plate)

27. A dirty tungsten electrode indicates that during a previous use all but which one of the following occurred?
    a. the gas was shut off before the electrode was cooled
    b. there was an air leak in the gas supply system
    c. the tip was contaminated by touching the weld pool
    d. the welding tip was oxidized

27. _____

28. When TIG welding with a current of 75 to 200 amperes, a welding operator should wear shaded lenses of glass number ____.
    a. 6   b. 10   c. 12   d. 14

28. _____

29. For TIG welding with a.c. high frequency current, the electrode does not have to touch the work to start the arc. (True or False)

29. _____

30. Which of the following is not one of the three arc-starting methods generally used for TIG welding?
    a. high frequency starting
    b. scratch starting
    c. low-voltage touch starting
    d. lift arc starting

30. _____

31. During TIG welding, arc blow may be caused by all but which one of the following?
    a. current set too high
    b. a contaminated electrode
    c. magnetic effects
    d. drafts in the work area

31. _____

32. During TIG welding, it is important to maintain a ____ arc.

32. _____

33. The extension of the tungsten electrode beyond the gas cup is governed by the shape of the object and the type of ____.

33. _____

34. Sharpen the electrode end to a fine point in all but which one of the following?
    a. welding light-gauge carbon steel
    b. welding aluminum
    c. concentrating heat on a limited area
    d. welding stainless steel

34. _____

35. If the addition of filler rod is necessary, choose a rod size greater than the thickness of the metal being welded. (True or False)

35. _____

NAME: _____   DATE: _____

36. During TIG welding, the longer the extension of the electrode beyond the cup, the less effective the gas shielding. (True or False)

36. _____

37. To finish a TIG weld, the arc should be broken abruptly. (True of False)

37. _____

38. A gas ____ can be used where a long electrode extension is required to aid in getting proper gas coverage.

38. _____

39. Of the metals most commonly TIG welded, the easiest to weld is ____.
   a. aluminum   b. carbon steel
   c. low alloy steel   d. stainless steel

39. _____

40. During TIG welding, an erratic arc may occur in all but which one of the following?
   a. the electrode diameter is too large
   b. the arc is too long
   c. the joint is too wide
   d. the base metal is dirty

40. _____

41. During TIG welding, the hot filler rod should be within the shielding gas area to prevent excessive oxidation. (True or False)

41. _____

42. To control penetration and weld contour during TIG welding, use an arc length ____.
   a. equal to the electrode diameter
   b. smaller than the electrode diameter
   c. a little longer than the width of the joint
   d. a little shorter than the width of the joint

42. _____

43. For TIG welding, always use the largest gas cup possible for a given weld. (True or False)

43. _____

44. Alternating the addition of filler metal to the weld pool with torch movement makes a better appearing bead. (True or False)

44. _____

45. The easiest joint to weld with the TIG process is the ____.
   a. butt   b. T   c. lap   d. corner   e. edge

45. _____

46. Filler rods must be added to TIG-welded ____.
   a. edge joints   b. lap joints
   c. T-joints   d. butt joints

46. _____

47. The center of the weld pool for TIG-welded lap joints is called the ____.

48. Multipass fillet welding with the TIG process is generally necessary when the material thickness is more than ____.
   a. 1/8"   b. 3/16"   c. 5/32"   d. 1/4"

49. The single-V butt joint is used when complete penetration is required on material thicknesses ranging between ____.

50. In a stainless steel but weld, a dark purplish-blue bead with hardly noticeable ripples indicates excessive ____.

47. _____

48. _____

49. _____

50. _____

Score _____
(50 possible)

NAME: _____ DATE: _____

CHAPTER 20

Gas Tungsten-Arc Welding Practice (Pipe)

Choose the letter of the correct answer. If there is no lettered answer, write the word(s) answer.

1. The TIG welding process can be used for welding ferrous and nonferrous piping materials. (True or False)

   1. _____

2. Aluminum is preferred for applications requiring all but which one of the following?
   a. light weight
   b. corrosion resistance
   c. temperatures above 750° F.
   d. nontoxic materials for food processing

   2. _____

3. Metals especially resistant to corrosive chemicals include all but which one of the following?
   a. chromium   b. stainless steel
   c. copper   d. nickel

   3. _____

4. For welding carbon steel pipe, the TIG process is used only for the ____ pass.

   4. _____

5. The process of passing argon or helium through the pipe to shield the weld pool from the inside is called ____.

   5. _____

6. The inside of a pipe may be protected by painting it with an appropriate ____.

   6. _____

7. The formation of carbon monoxide or carbon dioxide during steel welding will cause ____ in the weld.

   7. _____

8. Porosity may result from the presence on a weld joint of all but which one of the following?
   a. rust   b. flux   c. oil
   d. grease   e. moisture

   8. _____

9. Porosity caused by inadequate gas shielding may be the result of all but which one of the following?
   a. gas flow that is too slow
   b. an electrode that is too large
   c. an arc that is too long
   d. air currents in the work area

   9. _____

CHAPTER 20: Gas Tungsten-Arc Welding Practice (Pipe) | *107*

10. When TIG welding pipe, the presence of a notch in the weld pool indicates the absence of penetration on the inside of the pipe. (True or False)

10. _____

11. Add filler rod to the weld puddle to correct ____ penetration in the root pass of a TIG weld on pipe.

11. _____

12. Walking the cup is a method that allows for more control on the ____ pass.

12. _____

13. Control of ____ is the most important factor in successful root-pass welding.

13. _____

14. To terminate a weld bead, direct the weld puddle off to the side of the joint. (True or False)

14. _____

15. Filler passes are used to fill the pipe joint to within ____ inch of the top surface of the pipe.

15. _____

16. Filler passes may be made using the stringer bead technique or the weaved bead technique. (True or False)

16. _____

17. Finished cover passes should include all but which one of the following?
   a. be 3/32" wider than the beveled edges on each side of the joint
   b. have no overlap
   c. be slightly convex
   d. have a crown about 1/16" above the pipe surface

17. _____

18. Pipe joints welded in the vertical fixed position should be made with ____ beads.

18. _____

19. TIG welding pipe in the vertical fixed position requires special techniques to compensate for ____ of the weld puddle caused by gravity.

19. _____

20. Welding pipe in the horizontal fixed position enables a welder to practice three welding positions. (True or False)

20. _____

21. When TIG welding pipe in the horizontal fixed position, the weld bead should be started at the bottom center. (True or False)

21. _____

22. Downhill travel does not produce satisfactory results when pipe is being TIG welded. (True or False)

22. _____

NAME: _____ DATE: _____

23. When welding pipe with the TIG process, the second pass must be applied with great care because the root pass is likely to be ____.

23. _____

24. Soundness tests for TIG pipe welds are similar to those for plate. (True or False)

24. _____

25. Lateral joints may also be referred to as ____ joints.

25. _____

Score _____
(25 possible)

NAME: _____ DATE: _____

CHAPTER 21

Gas Metal Arc and Flux Cored Arc Welding Principles

Choose the letter of the correct answer. If there is no lettered answer, write the word(s) answer.

1. Gas metal-arc welding is a continuous ____ process.

2. The constant current welding machine is used for MIG welding. (True or False)

3. The use of the MIG process was extended by the introduction as a shield gas of ____.
   a. carbon dioxide   b. argon   c. hydrogen
   d. helium   e. oxygen

4. Some forms of MIG welding involve the use of flux. (True or False)

5. Match the types of open arc transfer to their best descriptions.

   1. Spray   a. Current alternates between high and low ranges.

   2. Globular   b. Transfer occurs below the surface of the base metal.

   3. Pulsed arc   c. Molten ball twice the diameter of the filler wire is transferred.

   4. Submerged arc d. Fine droplets of metal move rapidly from electrode to workpiece.

6. In short-circuiting arc transfer, no metal is transferred across the welding arc. (True or False)

7. Use only argon and helium mixtures as shielding gases for short circuit transfers. (True or False)

8. Instantaneous, flawless arc starts are characteristic of constant ____ machines.

1. _____

2. _____

3. _____

4. _____

5a. _____

5b. _____

5c. _____

5d. _____

6. _____

7. _____

8. _____

9. In MIG welding the amount of current is determined by the ____ at which the electrode wire is fed to the work.

9. _____

10. The constant voltage power source is designed to keep the wire-feed rate and the ____ rate equal.

10. _____

11. Arc voltage during MIG welding is higher than open circuit voltage. (True or False)

11. _____

12. The controls commonly found on constant voltage machines regulate all but which one of the following?
    a. voltage   b. current   c. slope   d. inductance

12. _____

13. ____ is the force that causes current to flow but does not flow itself.

13. _____

14. Most MIG welding is done using ____.
    a. a.c. straight polarity   b. d.c. straight polarity
    c. d.c. reverse polarity   d. a.c. reverse polarity

14. _____

15. Most wire feeders ____ the wire through the cable to the torch.

15. _____

16. Wire feeders that pull wire through the cable to the torch use wire that is all but which one of the following?
    a. hard   b. soft   c. aluminum   d. manganese

16. _____

17. Shielding gas nozzles must be large enough to allow proper gas flow yet small enough to allow access to the ____.

17. _____

18. All MIG welding torches have a ____ switch that turns on controls for wire feed shielding gas and water flow.

18. _____

19. The shielding gas system supplies and controls the flow of gas which shields the arc area from the surrounding ____.

19. _____

20. Gas from the cylinder is controlled by a pressure-reducing regulator solenoid control valves and a ____.

20. _____

21. To provide a good gas shield, ____ flow must be established.

21. _____

22. A gas flow rate that is too high causes ____.

22. _____

23. Of the following, the least dense is ____.
    a. air   b. carbon dioxide   c. helium   d. argon

23. _____

NAME: _____ DATE: _____

24. Denser gases require higher flow rates for adequate shielding than do lighter gases. (True or False)

24. _____

25. Such inert gases as xenon, krypton, and neon are seldom used for MIG welding. (True or False)

25. _____

26. To stabilize the welding arc, use ____ alone or mixed with another shielding gas.

26. _____

27. When used under appropriate conditions, argon shielding produces a narrow weld with deep penetration at the center. (True or False)

27. _____

28. Argon is preferred for welding heavy gauge metals. (True or False)

28. _____

29. Argon should never be used for out-of-position welding. (True or False)

29. _____

30. Helium shielding requires a ____ gas flow rate than argon shielding.

30. _____

31. Helium shielding produces a ____, slightly less penetrating weld than argon shielding.

31. _____

32. Helium is seldom used with automatic and mechanized processes. (True or False)

32. _____

33. Carbon dioxide gas is all but which one of the following?
    a. not inert
    b. composed of carbon and oxygen
    c. a contributor to weld spatter
    d. used for deep penetration
    e. known to cause an unstable arc

33. _____

34. Welding with carbon dioxide produces poisonous gas. (True or False)

34. _____

35. The presence of even a little moisture in carbon dioxide gas affects the quality of the weld. (True or False)

35. _____

36. Shielding gases that are inert may be mixed with shielding gases that are not. (True or False)

36. _____

CHAPTER 21: Gas Metal Arc and Flux Cored Arc Welding Principles

37. Materials that require filler wire of composition different than the base metal are all but which one of the following?
    a. copper alloys
    b. zinc alloys
    c. magnesium alloys
    d. high strength aluminum
    e. high strength steel alloys

38. Match the welds in Fig. 21-1 to the shielding gases used to make them.

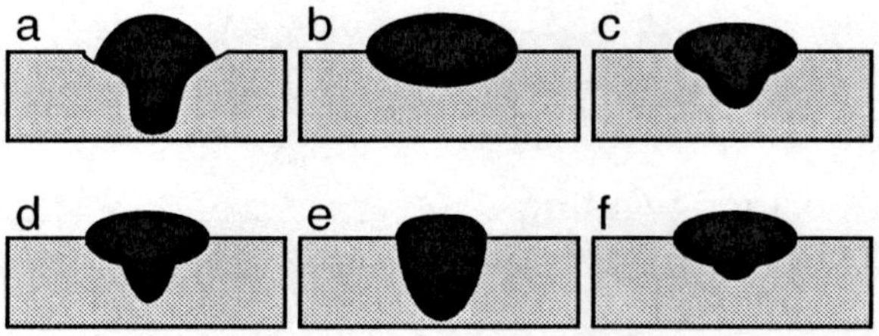

FIG. 21-1.

1. Helium/Argon

2. Argon/CO_2

3. Argon

4. CO_2

5. Helium

6. Argon/Oxygen

39. Filler wires as large as 1/4 inch in diameter are available for MIG welding. (True or False)

40. To prevent weld porosity, deoxidizers are used as scavenging agents. (True or False)

41. The T in the AWS classification for flux-cored filler wires stands for ____.

42. The flux is the primary method of carrying the ____ and deoxidizing elements to the weld pool in a flux cored electrode.

40. _____

38a. _____

38b. _____

38c. _____

38d. _____

38e. _____

38f. _____

39. _____

40. _____

41. _____

42. _____

Score _____
(50 possible)

NAME: _____ DATE: _____

CHAPTER 22

Gas Metal-Arc Welding Practice With Solid and Metal Core Wire (Plate)

Choose the letter of the correct answer. If there is no lettered answer, write the word(s) answer.

1. ____ current is generally used for GMAW.

2. For GMAW, ____ or carbon dioxide may be added to the shielding gas to reduce porosity and improve arc stability and weld appearance.

3. Argon shielding gas gives better GMAW coverage than helium because it is ____ than air.

4. The use of carbon dioxide for GMAW has all of the following advantages except which one?
   a. low cost
   b. low density
   c. low flow rates
   d. less burn

5. Pure argon is the preferred shielding gas for all but which one of the following?
   a. titanium
   b. magnesium
   c. nickel
   d. stainless steel

6. Add ____ to argon to increase penetration.

7. Joint designs like those used for other welding processes can be used for GMAW with a reduction in costs. (True or False)

8. Compared to shielded metal-arc welding, GMAW is all but which one of the following?
   a. is more penetrating
   b. allows wider groove-plate openings
   c. is faster
   d. requires less weld metal

1. _____

2. _____

3. _____

4. _____

5. _____

6. _____

7. _____

8. _____

9. The position of the welding gun in respect to the joint is expressed in terms of two angles that may be identified by all the following terms except which one?
   a. drag
   b. nozzle
   c. longitudinal
   d. transverse

9. _____

10. During GMAW the drag angle technique produces deeper penetration than the push angle technique. (True or False)

10. _____

11. When welding a T-joint with GMAW, the ____ angle can be used to prevent undercut.

11. _____

12. Using arc voltage that is too high for a given GMAW current does all but which one of the following?
    a. causes globular transfer of weld metal
    b. increases porosity
    c. decreases bead height and width
    d. reduces burn-off rates

12. _____

13. During GMAW, high travel speeds do all but which one of the following?
    a. produce large weld beads
    b. provide shallow penetration
    c. introduce less heat to the work
    d. may cause spatter

13. _____

14. For GMAW, the operating variables are adjusted on the basis of all but which one of the following?
    a. the type of material being welded
    b. the thickness of the material
    c. the position of welding
    d. type of welding machine used

14. _____

15. If GMAW wire-feed speeds are excessive, ____ or stubbing results because there is not enough current to melt the wire fast enough.

15. _____

16. If a given GMAW travel speed and wire-feed speed are not providing sufficient weld metal, all but which one of the following occur.
    a. increase both
    b. increase travel or decrease wire feed
    c. decrease travel or increase wire feed
    d. decrease both

16. _____

116 | STUDENT WORKBOOK

NAME: _____   DATE: _____

17. Match the following GMAW discontinuities to the combination of procedures most likely to correct them.

1. Incomplete penetration
2. Excessive penetration
3. Whiskers
4. Voids
5. Incomplete fusion
6. Porosity
7. Warpage
8. Cracking
9. Excessive spatter
10. Irregular weld shape
11. Undercutting

a. Use shield gas properly; clean and maintain equipment; reduce current
b. Reduce current or increase travel speed.
c. Control expansion and contraction.
d. Weave welding gun; hesitate at each side of the joint.
e. Increase current; reduce stickout distance.
f. Use compatible filler and base metals; allow for expansion and contraction.
g. Handle torch properly; use appropriate shield gas; maintain proper travel speed, current, arc voltage, and stickout.
h. Reduce travel and wire-feed speeds; increase stickout; weave gun.
i. Cover all joint areas with arc; reduce pool size; lead pool with electrode; check current values.
j. Reduce current; shorten and stabilize arc; clean nozzle; use appropriate shielding gas.
k. Fill all passes; increase arc voltage and travel speed following defective passes.

17a. _____
17b. _____
17c. _____
17d. _____
17e. _____
17f. _____
17g. _____
17h. _____
17i. _____
17j. _____
17k. _____

CHAPTER 22: Gas Metal-Arc Welding Practice with Solid Core Wire (Plate)

18. For the radiation produced by GMAW all but which one of the following is true?
    a. is the highest produced by any welding process
    b. increases with the use of argon
    c. disintegrates cotton clothing
    d. increases spatter from hot metal and slag

18. _____

19. For maximum eye protection during GMAW all but which one of the following is correct?
    a. wear very dark helmet glass
    b. always shield the eyes when less than 20 feet from the arc
    c. wear safety glasses under the welding helmet
    d. use a heavier lens shade for indoor welding

19. _____

20. Special ventilation precautions should be taken when GMAW under all but which one of the following?
    a. metals that emit toxic fumes
    b. near degreasers
    c. using carbon dioxide shielding gas
    d. using nitrogen shielding gas
    e. in the presence of any amount of ozone

20. _____

21. Electrical hazards are not quite as great with GMAW as with other processes. (True or False)

21. _____

22. Whenever possible, GMAW should be done in the ____ position to increase penetration and deposition rates.

22. _____

23. MIG welding requires the use of ____ current welding machines.

23. _____

24. The current value on a MIG welding machine can be checked on the machine ammeter when the welder is not welding. (True or False)

24. _____

25. Arc voltage controls all but which one of the following?
    a. penetration   b. bead contour
    c. undercutting   d. bead size   e. porosity

25. _____

26. GMAW is more expensive than other welding processes. (True or False)

26. _____

27. Fillet welds made by the GMAW process may be smaller than those made with stick electrodes. (True or False)

27. _____

28. The GMAW process is generally used for welding ____ gauges of aluminum.

28. _____

NAME: _____ DATE: _____

29. Minimize arc blow by welding away from the work connection. (True or False)

29. _____

30. With GMAW the current level is adjusted using the ____ knob.

30. _____

31. Aluminum ____ joints perform better under fatigue loading than other kinds of joints.

31. _____

32. Porosity in aluminum GMAW welds is commonly caused by all but which one of the following?
   a. hydrogen in the weld area
   b. a leaky gun
   c. improper joint preparation
   d. insufficient shielding gas

32. _____

33. GMAW welding stainless steel does all but which one of the following?
   a. reduces cleaning time
   b. makes the weld pool visible
   c. allows continuous wire feed
   d. uses only the fast, spray arc technique

33. _____

34. All but which one of the following metals are readily welded by the GMAW process?
   a titanium   b. Monel   c. electrolytic copper
   d. magnesium   e. Inconel

34. _____

35. Argon is recommended for GMAW of magnesium because of its excellent ____ action.

35. _____

36. Lead, phosphorus, and sulfur improve nickel alloys' resistance to cracking when heated. (True or False)

36. _____

37. All but which one of the following is true of titanium?
   a. is the only element that burns in nitrogen
   b. melts at a low temperature
   c. is used in electrode coatings
   d. is used as a stabilizer in stainless steel

37. _____

38. All but which one of the following is true of titanium?
   a. remains solid up to 4,500° F.
   b. scratches glass
   c. is used in hard-facing material
   d is usually used in the pure state

38. _____

CHAPTER 22: Gas Metal-Arc Welding Practice with Solid Core Wire (Plate)

39. Special shielding precautions are necessary for gas metal-arc welding titanium and zirconium. (True or False)

39. _____

40. ____ is the preferred shielding gas for welding aluminum plate up to 1 inch.

40. _____

Score _____
(50 possible)

NAME: _____ DATE: _____

CHAPTER 23

Flux-Cored Arc Welding Practice (Plate), Submerged Arc Welding, and Related Processes

Choose the letter of the correct answer. If there is no lettered answer, write the word(s) answer.

1. Flux-cored arc welding increases welding speeds and ____ rates.

2. Flux-cored arc may be gas or ____ shielded.

3. Both types of flux cored arc welding can produce welds of the highest ____.

4. FCAW-S electrodes have deposition efficiencies up to 90 percent; where FCAW-S electrodes have less deposition efficiencies, up to 87 percent. (True or False)

5. Direct current electrode ____ is most commonly used for FCAW-S electrodes.

6. FCAW-G electrodes are used with such shielding gases as all but which one of the following?
    a. carbon dioxide
    b. nitrogen
    c. argon plus carbon dioxide

7. All but which one of the following flux-cored filler electrodes may be used with external shielding gas?
    a. E70T-1   b. E70T-2   c. E70T-3   d. E70T-9

8. All but which one of the following flux-cored filler electrodes require no external shielding gas?
    a. E70T-3   b. E71T-7   c. E70T-10   d. E70T-12

9. Because flux cored electrodes carry an AWS Classification number, it is acceptable to switch between the various manufacturers. (True or False)

1. _____

2. _____

3. _____

4. _____

5. _____

6. _____

7. _____

8. _____

9. _____

10. M Type electrodes require ____% argon/balance CO2: if M is not present CO2 is to be used.

11. It may be necessary to ____ gas-flow rates when flux-cored electrodes welding is done outside.

12. When welding with flux-cored arc welding process, a push (forehand) nozzle angle does all but which one of the following?
    a. directs gas shield ahead of weld puddle
    b. plays the arc stream on cold base metal
    c. produces a narrow weld bead,
    d. reduces burn-through on thin plate

13. Gas-shielded, flux-cored arc welding is usually done with a minimum extension in inches of ____.
    a. 1/8"   b. 1/4"   c. 1/2"   d. 1"

14. Self-shielded, flux-cored arc welding is usually done with a minimum extension in inches of ____.
    a. 1/8"   b. 1/4"   c. 1/2"   d. 1"

15. Electrode extension that is too long results in all but which one of the following?
    a. excessive spatter   b. irregular arc action
    c. loss of gas shielding d. excessive penetration

16. Gas-shielded, flux-cored arc welding deposits are lower in hydrogen than those made with low hydrogen stick electrodes. (True or False)

17. During gas-shielded, flux-cored arc welding, all but one of the following are true. Which one is incorrect?
    a. too long an arc causes spatter
    b. cleaning the plate reduces undercut
    c. penetration decreases as travel speed increases
    d. drag welding produces deeper penetration

18. When welding with gas-shielded, flux-cored arc electrodes, travel speed that is too fast for the current produces the all but which one of the following weld characteristics?
    a. slag entrapment   b. undersized bead
    c. shallow penetration   d. undercutting

10. _____

11. _____

12. _____

13. _____

14. _____

15. _____

16. _____

17. _____

18. _____

NAME: _____ DATE: _____

19. The equipment needed for self-shielded, flux-cored arc welding includes all but which one of the following?
    a. constant voltage power source
    b. continuous wire-feed mechanism
    c. auxiliary gas shielding

19. _____

20. The self-shielded, flux-cored arc contains all the necessary ingredients for shielding, deoxidizing, ____ and alloying materials.

20. _____

21. The self-shielded, flux-cored wire process offers all but which one of the following advantages?
    a. all-position welding
    b. faster, more flexible installation
    c. heavier, more stable welding guns
    d. uninterrupted wire feeding

21. _____

22. Constant voltage, d.c. power source are used for welding FCAW-S process and electrodes. These machines can be all but which one of the following types?
    a. transformer rectifier type
    b. inverter type
    c. semi-automatic type
    d. engine-driven generator type

22. _____

23. Only about 1/2 of the total extension length is visible beyond the end of the nozzle on the welding gun. (True or False)

23. _____

24. Increasing the arc voltage during self-shielded, flux-cored arc welding need not affect the other three welding variables. (True or False)

24. _____

25. Magnetic-flux coated wire welding requires all but which one of the following?
    a. coated steel welding wire
    b. magnetic flux
    c. carbon dioxide shielding gas
    d. a power supply

25. _____

26. Submerged-arc welding takes place beneath a blanket of granular, fusible ____ which completely hides the arc.

26. _____

CHAPTER 23: Flux-Cored Arc Welding Practice and Submerged Arc Welding

27. For submerged-arc welding, which one of the following is not correct?
    a. the filler wire lightly touches the workpiece
    b. there is no spatter, smoke, or sparks
    c. goggles are the only protective equipment needed
    d. the process may be automatic or semi-automatic

27. _____

28. Some smoke and fumes are created that require proper ____.

28. _____

29. Submerged-arc welds have all but which one of the following characteristics?
    a. high impact strength
    b. uniformity
    c. density
    d. corrosion resistance
    e. high nitrogen content

29. _____

30. Submerged-arc welding must be done in the ____ position.

30. _____

31. For alloy pipe welding with the submerged-arc process, the alloying elements should be mixed with the flux. (True or False)

31. _____

32. The power source for submerged-arc welding may be AC, DCEP or DCEN. (True or False)

32. _____

33. ____ provides the best control of bead shape and maximum penetration.

33. _____

34. For constant voltage welding machines, which one of the following is not correct?
    a. use constant-feed wire control
    b. is preferred for small diameter filler wire
    c. are recommended for hard-surfacing
    d. are preferred for high-speed welding of think materials

34. _____

35. For the fluxes used for submerged-arc welding, which one of the following is not true?
    a. carry high welding currents
    b. clean the base metal
    c. add no new properties to the weld metal
    d. provide maximum resistance to cracking

35. _____

124 | STUDENT WORKBOOK

NAME: _____ DATE: _____

36. The fluxes used with the submerged-arc welding process contain all but which one of the following?
   a. silicon   b. chromium   c. manganese
   d. aluminum   e. deoxidizers

36. _____

37. The largest diameter filler wire used with the submerged-arc welding process is 1/4 inch. (True or False)

37. _____

38. DCRP is recommended for tandem arc welding applications. (True or False)

38. _____

39. Heavy, deep-groove, submerged-arc welding is often done slightly down hill. (True or False)

39. _____

40. Joints with gaps greater than 1/16" may be filled with SMAW, GMAW, or FCAW process. (True or False)

40. _____

41. When welding long seams on a tank, eliminate cracking by tack welding steel ____ at each seam end.

41. _____

42. Excessive flux produces a narrow hump bend. (True or False)

42. _____

43. For submerged-arc welding, a long extension reduces costs by increasing the deposition rate and the welding speed. (True or False)

43. _____

44. During submerged-arc welding at a fixed current setting, increasing the electrode size affects the ____ of penetration.

44. _____

45. For multiple arcs, which one of the following is not correct?
   a. increase meltoff   b. increase speed
   c. use one power source   d. reduce deposition rates.

45. _____

46. The SAW process is often used to ____ carbon steel with stainless steel.

46. _____

47. Semi-automatic, submerged-arc welding is a ____ wire process recommended for irregular shapes.

47. _____

48. A submerged-arc process developed especially for single-pass welding of thick vertical plates is ____ welding.

48. _____

CHAPTER 23: Flux-Cored Arc Welding Practice and Submerged Arc Welding | 125

49. For semi-automatic, submerged-arc welding, which one of the following is not correct?
    a. requires no plate edge preparation
    b. is a manual process
    c. smooth, spatter free bead
    d. is a high-heat process

49. _____

50. Consumable-guide electroslag welding is not a true submerged-arc process. (True or False)

50. _____

51. EGW can be done with solid metal cored or flux cored electrodes. (True or False)

51. _____

52. During electrogas welding, the weld pool is protected by shielding gas. (True or False)

52. _____

53. Electrogas welding uses direct current electrode positive. (True or False)

53. _____

54. The selection the welding process depends on the proper evaluation of each job. (True or False)

54. _____

55. Automatic welding processes are ideal for ____.
    a. short welds   b. complicated shapes
    c. repetitive, fixtures jobs   d. heavier construction

55. _____

Score _____
(55 possible)

NAME: _____ DATE: _____

CHAPTER 24

Gas Metal Arc Welding Practice (Pipe)

Choose the letter of the correct answer. If there is no lettered answer, write the word(s) answer.

1. For MIG welding pipe which one of the following is not true?
   a. is often faster than other processes
   b. increases cleaning time
   c. may not require backup rings
   d. reduces warpage and distortion

   1. _____

2. A d.c. constant current welding machine is used for the MIG welding of pipe. (True or False)

   2. _____

3. For MIG welding, a quiet power source that draws current only during welding is ____.
   a. a motor-driven generator
   b. a d.c. rectifier
   c. an engine-driven generator
   d. an a.c.-d.c. transformer-rectifier

   3. _____

4. Power supplies for MIG welding should have all but which one of the following?
   a. variable slope control
   b. variable inductance control
   c. a hot start feature
   d. a special shut-off control

   4. _____

5. A gas flow rate of ____ cubic feet per hour is adequate for most indoor welding applications.

   5. _____

6. For spray transfer GMAW, the minimum argon content is ____.
   a. 70%   b. 80%   c. 90%   d. 100%

   6. _____

7. A tri-mix gas for GMAW of stainless steel would consist of all but which one of the following gases?
   a. helium   b. nitrogen   c. argon   d. CO_2

   7. _____

8. GMAW is considered to be a low hydrogen process. (True or False)

   8. _____

CHAPTER 24: Gas Metal Arc Welding Practice (Pipe) | *127*

9. For best results, pipe should be welded with E70S-3 filler wire with a diameter of ____ inch.

9. _____

10. Using a smaller filler wire for MIG welding at a given current level requires decreasing the wire feed speed. (True or False)

10. _____

11. Match the common MIG welding discontinuities to the appropriate remedies.

   1. Convex bead   a. Clean weld thoroughly.

   2. Scattered porosity   b. Increase root opening and voltage.

   3. Cold lapping   c. Speed up travel, decrease root opening.

   4. Lack of root fusion   d. Raise voltage; widen oscillation.

   5. Suckback in overhead position   e. Lower wire-feed speed; increase forward speed; keep arc ahead of puddle.

11a. _____

11b. _____

11c. _____

11d. _____

11e. _____

12. To remove tack defects and to insure good fusion to tack welds for MIG welding, both ends of all tacks should be ____.

12. _____

13. If a backing ring is used for MIG welding, a root opening of at least ____ inch is necessary.

13. _____

14. The ____ joint is the most commonly used pipe joint in welded pipe systems.

14. _____

15. A 37 1/2-degree bevel and uphill travel are used for MIG welding ____ piping.

15. _____

16. Cross-country and ____ piping have a 30-degree bevel and are welded downhill.

16. _____

17. When MIG welding a rolled butt weld in pipe, start the first filler pass about ____ inches past the original starting point for the root pass.

17. _____

NAME: _____ DATE: _____

18. The bell-hole position is another name for ____. 18. _____
   a. horizontal roll position
   b. vertical fixed position
   c. horizontal fixed position
   d. vertical roll position

19. It is important to keep the arc ____ of the weld pool 19. _____
    when doing filler and cover passes down hill.

20. Fillet and groove welds are used for ____ joints in pipe. 20. _____

21. The heel of the joint referred to in question 24 should be 21. _____
    welded first. (True or False)

Score _____
(25 possible)

NAME: _____ DATE: _____

CHAPTER 25

High Energy Beams and Related Welding and Cutting Process Principles

Choose the letter of the correct answer. If there is no lettered answer, write the word(s) answer.

1. High energy beams are concentrated heat sources that have measured as high as ____ watts/square inch.

2. All normal type joints can be welded with high energy beam processes. (True or False)

3. High energy beams generally produce very narrow welds with very deep ____.

4. Complete joint penetration (CJP) welds using the keyhole technique require joint tolerances in the range of ____.
   a. +/-0.0001"   b. +/-0.001"
   c. +/-0.010"   d. +/-0.100"

5. Which high-energy process is accomplished by the use of a concentrated stream of high-velocity electrons formed into a beam?

6. Laser beam welding (LBW) is generally done with an ____ gas to shield the weld pool.

7. Lasers can cut non-metals, such as plastic wood or cloth. (True or False)

8. ____ are unwanted molten metal flying out of the cut, interrupting the cut path.

9. When laser beam cutting, an ____ gas is used to help improve combustion and physically blow metal from the kerf.

10. The assist gas recommended for laser beam cutting of copper is ____.
    a. air   b. oxygen   c. nitrogen   d. argon

1. _____

2. _____

3. _____

4. _____

5. _____

6. _____

7. _____

8. _____

9. _____

10. _____

11. Which one of the following is not an advantage of a laser beam?
    a. very low heat input
    b. non-contact process
    c. very few metals can be processed
    d. very small welds and holes can be made

12. Water jet cutting uses a high-velocity jet of water at pressures of up to _____ psi.

13. Which of the following materials is not used for the orifice of a water jet cutting nozzle?
    a. jade   b. diamond   c. sapphire   d. ruby

14. Water jet cutting produces narrow kerfs of _____ inch.

15. How fast can Abrasive Water Jet cut 3/4" Armor plate steel?

16. The welding process where a probe or tap with a diameter of 0.20" to 0.24" is rotated between the square groove faying edges on a butt joint is _____.
    a. laser beam   b. electron beam
    c. forge   d. friction stir

17. A solid state process that uses a controlled detonation to impact two work pieces at a very high velocity is _____.

18. A solid state process that uses the heat produced by compressive forces generated by materials rotating together is _____.

19. Laser assisted arc welding is a hybrid process using a laser in combination with _____.
    a. SMAW   b. SAW   c. FCAW   d. GMAW

20. The American Welding Society has defined and described over 110 various joining and cutting processes and process variations. (True or False)

11. _____
12. _____
13. _____
14. _____
15. _____
16. _____
17. _____
18. _____
19. _____
20. _____

Score _____
(20 possible)

NAME: _____ DATE: _____

CHAPTER 26

General Equipment for Welding Shops

Choose the letter of the correct answer. If there is no lettered answer, write the word(s) answer.

1. In shop areas designed specifically for welding, permanent ____ are erected to protect other welders.

    1. _____

2. The use of positioners to enable welding in the flat position has done all but which one of the following?
    a. decreased production
    b. reduced costs
    c. improved quality
    d. promoted safety in both production and repair welding

    2. _____

3. Turning rolls are used to weld circumferential seams in ____.

    3. _____

4. Wheel-type turning rolls may be all of the following except which one?
    a. driver-idler units   b. unit-frame rolls
    c. chain-sling units   d. steel tired   e. rubber tired

    4. _____

5. A weld ____ has three movable jaws similar to the ones on a lathe chuck.

    5. _____

6. A ____ enables arc welding units to move vertically and horizontally as well as to rotate a full 360'.

    6. _____

7. A turntable is used for over head welding. (True or False)

    7. _____

8. There are four basic types of Seamers. (True or False)

    8. _____

9. A weld ____ raises or lowers the welder himself along a vertical surface.

    9. _____

10. There are a number of clamps and holding devices that employ magnetic attraction. (True or False)

    10. _____

11. For precision linear travel, a side beam carriage is an effective tool. (True or False)

12. There are many advantages to hydroforming including all but which one of the following?
    a. reduce tooling cost
    b. increase part weight
    c. reduce the number of weld joints
    d. increase part stiffness

13. An orbital welding machine is used to make fillet welds in all positions. (True or False)

14. The C clamps and wedges important to carpentry are merely excess baggage for welders. (True or False)

15. To be sure that holes in hot metal do not lose their shape, reinforce them with round ____ sticks.

16. Heat-treating ovens cannot be fired by electricity. (True or False)

17. Cleaning a surface for welding and removing scale, slag, and rust after welding, is done by ____.

18. Materials used as abrasives in sandblasting equipment include all of the following except which one?
    a. aluminum oxide   b. metal grit   c. lead filings
    d. walnut shells   e. corn cobs

19. Using heat and pressure to weld two lapped pieces is ____ welding.

20. In modern welding shops, hand tools have been replaced by ____ ones.

21. A versatile brake that can form four sides and a bottom from one flat metal sheet is a ____.
    a. box and pan brake   b. hand brake
    c. roll   d. power press brake

22. An indispensable piece of equipment for bending radii and angles on various shapes is a ____ bender.

23. Power squaring shears may be used on metal plate not more than 1/2-inch thick. (True or False)

11. _____

12. _____

13. _____

14. _____

15. _____

16. _____

17. _____

18. _____

19. _____

20. _____

21. _____

22. _____

23. _____

NAME: _____ DATE: _____

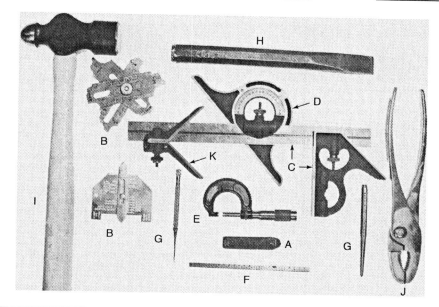

FIG. 26-1.

24. Match the tools in Fig. 26-1 to the appropriate names.
   1. Weld gauges
   2. Micrometer
   3. Pliers
   4. Cold chisel
   5. Identification stamp
   6. Center head
   7. Combination square and scale
   8. Hand hammer
   9. Scribe and center punch
   10. Bevel protractor
   11. Flexible scale

   24a. _____
   24b. _____
   24c. _____
   24d. _____
   24e. _____
   24f. _____
   24g. _____
   24h. _____
   24i. _____
   24j. _____
   24k. _____

25. Of the four types of portable power tools, universal electric ones are most common. (True or False)

   25. _____

26. For universal electric power tools all of the following are true except which one?
   a. use the cheapest power available
   b. are ideal for continuous operation
   c. require expensive maintenance
   d. meet the demands of field work

   26. _____

CHAPTER 26: General Equipment for Welding Shops | 135

27. Compressed air is required for the operation of ____ power tools.

27. _____

28. Pneumatic tools are suitable for continuous operation. (True or False)

28. _____

29. Hydraulic power tools are operated by hydraulic ____.

29. _____

30. Match the following power tools to their best descriptions.

    1. Electric hammers   a. Produce a smooth cutting edge without distorting the body of the metal.

30a. _____

    2. Grinders   b. Used for grinding, buffing, and finishing aircraft, missile, and industrial operations.

30b. _____

    3. Shears and nibblers   c. Used for chipping, peening, channeling, and masonry drilling.

30c. _____

    4. Magnetic-base drill press   d. Used for economic, precision edge preparation.

30d. _____

    5. Beveling machine   e. Attached to steel surfaces for difficult drilling jobs.

30e. _____

    6. Weld shaver   f. Abrasive surfacing tools.

30f. _____

31. Tools that cut metal fit into the general category of ____ tools.

31. _____

32. Match the machine tools to their best descriptions.

    1. Engine lathe   a. Multiple-tool cutting tool used to cut gear teeth.

32a. _____

    2. Shaper   b. Turning tool that revolves the metal workpiece against a cutting edge.

32b. _____

    3. Milling machine   c. Cutting tool for rapid puncture of burr-free holes in various sizes.

32c. _____

    4. Pedestal Grinder   d. Polishing and cutting tool with a vitrified, silicate, or elastic wheel.

32d. _____

    5. Power punch   e. Cutting tool whose chipping action removes excess weld from specimen face and root.

32e. _____

    6. Metal cutting band saw   f. Is used more than any other tool in the welding shop.

32f. _____

Score _____
(52 possible)

NAME: _____ DATE: _____

CHAPTER 27

Automatic and Robotic Arc Welding Equipment

Choose the letter of the correct answer. If there is no lettered answer, write the word(s) answer.

1. In The automatic and robotic method of applications, the welders or operators are required to control the key factors influencing the weld. (True or False)

   1. _____

2. In certain situations, it is more advantageous to move the arc and thus the weld pool than it is to require the motion control device to change the torch or gun angle. (True or False)

   2. _____

3. Magnetic arc controls are typically used with all but which one of the following processes.
   a. GMAW   b. GTAW   c. SAW   d. PAW

   3. _____

4. The devices in question 3 provide control over all but which one of the following?
   a. heat distribution   b. excessive porosity
   c. incomplete fusion   d. a wandering arc

   4. _____

5. Sophisticated mechanical oscillators are available such as the ____ controlled pendulum.

   5. _____

6. Arc monitoring does not require the welder or operator to be skilled or trained. (True or False)

   6. _____

7. There are many variables that must be dealt with that may adversely affect the completed weldment. (True or False)

   7. _____

8. There are four basic control elements that must be dealt with. Which one of the following is not one of the four?
   a. welding
   b. manipulation of the input variables
   c. what is happening with the disturbing input variables
   d. mission control

   8. _____

9. The manipulated input variables are those that directly affect the ____ response variables.

9. _____

10. Sensing devices are generally transducers that convert energy from one form to another. (True or False)

10. _____

11. ____ of the various types are of no value unless their data can be monitored and used to control the deviation that is taking place.

11. _____

12. Control over a welding operation for automatic or robotic applications is very involved; the fundamental goal is to deposit a satisfactory weld. (True or False)

12. _____

13. Having good control over sequence of operation of the weld or "weld sequence" is ____.

13. _____

14. Microprocessor Based Controllers allow digital readouts and accurate setting of the weld sequence. (True or False)

14. _____

15. Robotic arc welding systems are very flexible. (True or False)

15. _____

16. The two main types of robot arms are the ____ and the rectilinear.

16. _____

17. In 1997 NASA's robot lander, Pioneer, crawled across the surface of ____.

17. _____

18. The pool of professional welders is dwindling; in fact, ____ percent of the current welding workforce will be retiring in the next decade. This will leave a shortfall of skilled, knowledgeable welders.

18. _____

19. Robots do not always have the ability to return to the exact same position each time. (True or False)

19. _____

20. Robots are rated by how many inches a minute an axis can move in a second. (True or False)

20. _____

21. Occasionally a robot may have to be programmed for up to ____ axes of motion.

21. _____

22. There are ____ different levels of recommended robotic training for operations personnel.

22. _____

138 | STUDENT WORKBOOK

NAME: _____   DATE: _____

23. The AWS QC19 is a Standard for ____.   23. _____
    a. welder operations
    b. qualifications of robotic arc welding
    c. arc welding related disciplines
    d. certification of automated process technicians

24. Today automation can mean solutions from a single   24. _____
    robot to a full production line. (True or False)

25. It has been determined that only about ____ percent of   25. _____
    the small to medium sized manufacturing companies
    have installed robots.

                                                  Score _____
                                                       (25 possible)

NAME: _____ DATE: _____

CHAPTER 28

Joint Design, Testing, and Inspection

Choose the letter of the correct answer. If there is no lettered answer, write the word(s) answer.

1. Identify the five basic types of joints shown in Fig. 28-1.

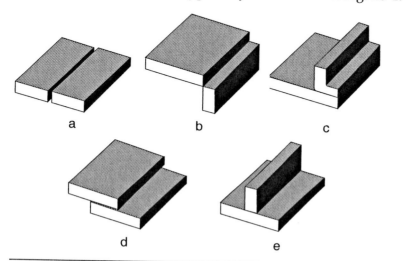

1a. _____

1b. _____

1c. _____

1d. _____

1e. _____

FIG. 28-1.

2. Open roots in joints ensure complete penetration in welding the joint. (True or False)

2. _____

3. ____ is the depth to which the base metal is melted and fused with the metal of the filler rod.

3. _____

4. Whether a joint should be set up as an open or closed root depends upon all but which one of the following?
   a. the thickness of the base metal
   b. the kind of joint
   c. the nature of the job
   d. the polarity of the electrode
   e. the position of welding

4. _____

5. Edge joints are both strong and economical. (True or False)

5. _____

6. Welding from both sides materially increases the strength of a joint. (True or False)

6. _____

CHAPTER 28: Joint Design, Testing, and Inspection | 141

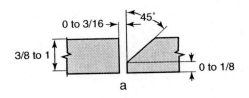
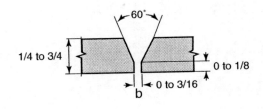

FIG. 28-2.

7. Identify the types of butt joints shown in Fig. 28-2.

7a. _____
7b. _____

8. Single-V butt joints provide for 100% penetration. (True or False)

8. _____

9. Root penetration is not necessary for joints that are beveled on both sides and have a root opening. (True or False)

9. _____

10. Match the types of corner joints to their best descriptions.
    1. Flush   a. Strong, butt fitup may be difficult.
    2. Half-open   b. Best used on light-gauge sheet metal.
    3. Full-open   c. Forms groove for root penetration.

10a. _____
10b. _____
10c. _____

11. Generally, product soundness and service life depend upon proper joint ____ and flawless welding.

11. _____

12. All but which one of the following are parts of a 3/4" butt weld?
    a. throat   b. heel   c. shoulder
    d. overlap   e. face

12. _____

13. Match the types of welds to their best descriptions.
    1. Strength   a. Extends the entire length of a joint.
    2. Caulking   b. Makes riveted joints leakproof.
    3. Composite   c. Holds job parts together in preparation for welding.
    4. Continuous   d. Carries the structural load.
    5. Intermittent   e. Placed at intervals to reduce cost of non-critical work.
    6. Tack   f. Meets load requirements and is also leakproof.

13a. _____
13b. _____
13c. _____
13d. _____
13e. _____
13f. _____

NAME: _____ DATE: _____

14. Match the welding positions to their descriptions.
    1. Flat   a. Weld travel may be up or down.
    2. Horizontal   b. Filler deposited from upper side of joint; weld face is horizontal.
    3. Vertical   c. Filler metal deposited from underside of joint; weld face is horizontal.
    4. Overhead   d. Filler is deposited on upper side of horizontal surface against a vertical surface.

14a. _____
14b. _____
14c. _____
14d. _____

15. Metal extending above the surfaces of the plate is called ____.

15. _____

16. High weld reinforcement increases the strength of a weld. (True or False)

16. _____

17. The width of a groove weld should not be more than ____ greater than the distance across the joint from edge to edge.

17. _____

18. The distance from the root to the toe of a fillet weld is called the ____.

18. _____

19. Throat thickness is the distance from the root to the ____ of a fillet weld.

19. _____

20. Visual inspection is an adequate means of determining the quality of a weld. (True or False)

20. _____

21. Different industries have different welding codes. (True or False)

21. _____

22. A welder must be thoroughly informed about the demands of existing welding codes. (True or False)

22. _____

23. The welding test that determines the correctness of the method of welding a specific project is the ____ qualification test.

23. _____

24. Tests that judge the welder rather than the work are ____ qualification tests.

24. _____

25. Evaluating some weld tests requires that the sample weld be destroyed beyond use. (True or False)

25. _____

CHAPTER 28: Joint Design, Testing, and Inspection | 143

26. The welding test most commonly used by the welder himself is ____.
    a. nondestructive   b. mechanical
    c. visual inspection   d. destructive

27. Magnetic particle testing can be used with all metals. (True or False)

28. The trade name by which magnetic particle testing is often called is ____.

29. Magnetic particle testing detects the presence of surface and internal cracks as much as ____ of an inch below the weld surface.

30. Oil, water, and slag from the welding operation do not interfere with magnetic particle testing. (True or False)

31. Radiographic inspection makes use of ____ rays and X-rays.

32. Radiographic inspection methods are used to show the presence and type of microscopic defects in the weld interior. (True or False)

33. Photographs for radiographic testing may be taken several feet away from work pieces that are hard to get close to. (True or False)

34. Penetrant inspection detects only ____ defects.

35. The fine white powder that soaks up the penetrant remaining on the workpiece to mark defects clearly is called a ____.

36. A penetrant test may also be conducted using a ____ material under special light instead of dyes and developers.

37. A closely controlled, rapid testing method that can probe deeply without damaging the weld is ____ inspection.

38. The testing method in question 19 is increasing in popularity because it is so easy to use. (True or False)

39. The testing method in question 19 is widely used by the ____ industry to detect defects and also to determine damage from corrosion and wear.

26. _____
27. _____
28. _____
29. _____
30. _____
31. _____
32. _____
33. _____
34. _____
35. _____
36. _____
37. _____
38. _____
39. _____

NAME: _____ DATE: _____

40. Eddy current testing may be used on both ferrous and nonferrous metals. (True or False)

40. _____

41. Leak tests apply pneumatic or hydraulic ____ to determine a weldment's resistance to leakage.

41. _____

42. Water can be used to detect leaks of all sizes. (True or False)

42. _____

43. Brinell and Rockwell are common nondestructive ____ tests.

43. _____

44. The test commonly used to determine the tensile strength of welds is ____.
   a. free-bend   b. transverse shear
   c. nick-break   d. reduced section tension

44. _____

45. Tests for weld soundness include all of the following except which one?
   a. longitudinal shear   b. root-bend
   c. nick-break   d. face-bend

45. _____

46. It is the general practice for code welding to qualify welders on the groove weld test in the 3G and 4G positions. (True or False)

46. _____

47. In making a butt weld for test purposes, it is important to keep the tensile strength of the plate and the weld metal about the same. (True or False)

47. _____

48. The most important part of a fillet test weld is the ____ pass.

48. _____

49. All grinding and machining of test specimens must always be lengthwise of the specimen. (True or False)

49. _____

50. A smooth finish of the test surface improves the chances of passing the test. (True or False)

50. _____

51. It is not necessary to remove bead reinforcement from the test weld surface. (True or False)

51. _____

52. No cracks or open defects of any kind are permitted on the surface of a test weld after it has been bent. (True or False)

52. _____

53. Longitudinal and transverse shear tests determine the shearing strength of ____ welds.

53. _____

CHAPTER 28: Joint Design, Testing, and Inspection

54. Etching reveals the penetration and soundness of a weld cross section. (True or False)

55. The Izod and Charpy tests are ____ tests.

56. ____ testing determines how well a weld can resist repetitive stress.

57. Match the principal weld defects to their best descriptions.

    1. Cracking
    2. Porosity
    3. Brittle welds
    4. Incomplete fusion
    5. Dimensional defects
    6. Incomplete penetration
    7. Inclusions
    8. Undercutting

    a. Failure of filler and base metal to fuse at the root of a joint
    b. Burning away of base metal at weld toe.
    c. Presence of inclusions containing gases rather than solids.
    d. Failure of layers of weld metal to fuse at any point in the weld groove.
    e. Contraction, warping, angular defects.
    f. Poor elongation, low-yield point, poor ductility, little resistance to stress.
    g. Presence of linear ruptures of metal under stress.
    h. Elongated or globular pockets of solid compounds.

58. ____ testing determines the tightness of welds in fabricated vessels.

59. Destructive testing gives an absolute measure of the strength of the sample tested. (True or False)

60. ____ testing is a good means of determining a welder's ability.

54. _____

55. _____

56. _____

57a. _____

57b. _____

57c. _____

57d. _____

57e. _____

57f. _____

57g. _____

57h. _____

58. _____

59. _____

60. _____

Score _____
(82 possible)

NAME: _____ DATE: _____

CHAPTER 29

Reading Shop Drawings

Choose the letter of the correct answer. If there is no lettered answer, write the word(s) answer.

1. Welders do not rely on print reading. It is the engineers' responsibilities. (True or False)

2. Assembly drawing show individual parts in detail. (True or False)

3. The term blueprint is used as often today as it was 25 years ago. (True or False)

4. Match the types of drawing lines to their best descriptions.

   1. Extension
   2. Center
   3. Hidden
   4. Cutting plane
   5. Leaders
   6. Break
   7. Projection

   a. Represent edges or surfaces that cannot be seen in the finished product.
   b. Point to a particular surface to show a dimension or a note.
   c. Extend away from an object to indicate its size.
   d. Show relationship of surfaces in one view with the same surfaces in other views.
   e. Indicate a break in an object or that only part of an object is shown.
   f. Sectional line to indicate where an imaginary cut is made.
   g. Dot-and-dash lines indicate the center of a circle or a part of a circle.

1. _____

2. _____

3. _____

4a. _____

4b. _____

4c. _____

4d. _____

4e. _____

4f. _____

4g. _____

5. Match the following terms to the appropriate drawing lines in Fig. 29-1.

   1. Phantom
   2. Cutting plane
   3. Dimension
   4. Border

   a
   b
   c
   d (6")

5a. _____

5b. _____

5c. _____

5d. _____

5. Leader — · — · — · —   5e. _____

6. Object – – – – – – –   5f. _____

7. Invisible edge — ·· — ·· —   5g. _____

8. Extension   5h. _____

9. Center   5i. _____

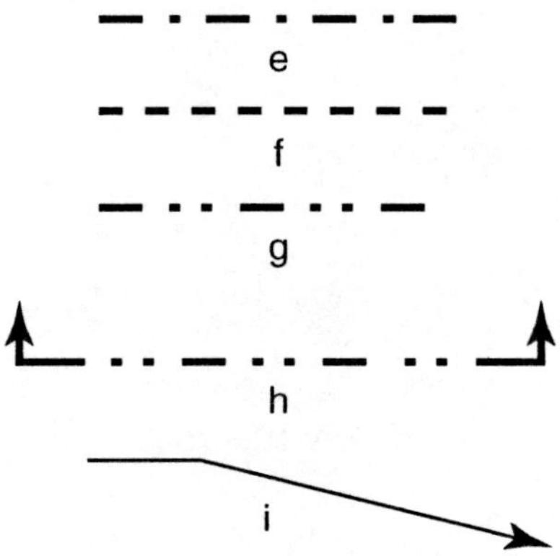

FIG. 29-1.

6. The metric measure of length is ____.   6. _____
   a. litre   b. hectare   c. gram   d. metre

7. A metric measure of area is ____.   7. _____
   a. litre   b. hectare   c. gram   d. metre

8. Volume may be measured in all but which one of the   8. _____
   following metric terms?
   a. cubic metres   b. litres
   c. centimetres   d. millilitres

9. Mass may be measured in all but which one of the   9. _____
   following metric terms?
   a. grams   b. tonnes
   c. square kilometres d. kilograms

10. In the answer column, write the metric terms indicated
    by the following symbols.
    a. kg   10a. _____
    b. ha   10b. _____
    c. m   10c. _____
    d. m3   10d. _____
    e. cm   10e. _____
    f. cm2   10f. _____
    g. t   10g. _____
    h. l   10h. _____

11. In the answer blank, write 20 degrees, 15 minutes, and 8   11. _____
    seconds using ____ the symbols for angular dimensions.

12. There are ____ degrees in a complete circle.   12. _____

148 | STUDENT WORKBOOK

NAME: _____ DATE: _____

13. The permissible range of variation in the dimensions of a completed job is called the limits of ____.

13. _____

14. Match the figures from the drawing of the lock keeper in Fig. 29-2 to the appropriate measurements.
    1. Overall length
    2. Thickness (in inches) of sheet brass.
    3. Recommended gauge thickness
    4. Diameter of drilled holes
    5. Number of drilled holes required
    6. Distance from the center of the bottom drilled hole to the bottom edge of the plate
    7. Distance from the center of the drilled holes to the left side of the keeper
    8. Length of the rectangular hole
    9. Width of the rectangular hole
    10. Distance from the bottom of the rectangular hole to the bottom edge of the keeper
    11. Distance from the top of the rectangular hole to the top edge of the keeper

    a. 2⅝"
    b. 0.064"
    c. 1/4
    d. 1 7/8"
    e. 1 1/4"
    f. 2
    g. 3/8"
    h. 5/8"
    i. 5 1/4"
    j. 14
    k. 9/16"

    14a. _____
    14b. _____
    14c. _____
    14d. _____
    14e. _____

    14f. _____
    14g. _____
    14h. _____
    14i. _____

    14j. _____

    14k. _____

Score _____
(45 possible)

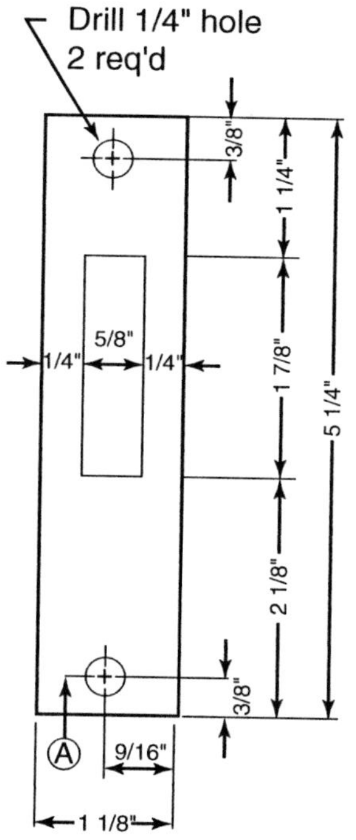

**LOCK KEEPER
SHEET BRASS
No. 14 (0.064") B. & S. GA.**

FIG. 29-2.

CHAPTER 29: Reading Shop Drawings | 149

NAME: _____   DATE: _____

CHAPTER 30

Welding Symbols

Choose the letter of the correct answer. If there is no lettered answer, write the word(s) answer.

1. The standard welding symbols have been developed by the ____.

2. Welding symbols give ____ welding information

3. Weld symbols are elements of welding symbols. (True or False)

4. Specifications for welding are given in the ____ of the reference line.

5. Match the following types of welds to the corresponding symbols, shown in Fig. 30-1.
   1. Plug or slot
   2. Edge
   3. V groove
   4. Fillet
   5. Seam
   6. Bevel groove
   7. Square groove

1. _____

2. _____

3. _____

4. _____

5a. _____

5b. _____

5c. _____

5d. _____

5e. _____

5f. _____

5g. _____

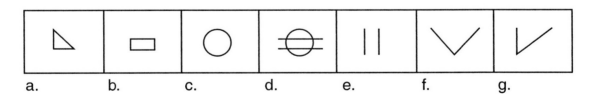

FIG. 30-1.

6. Match the following descriptions to the appropriate supplementary symbols shown in Fig. 30-2.
   1. Field weld
   2. All around weld
   3. Melt-through

6a. _____
6b. _____
6c. _____

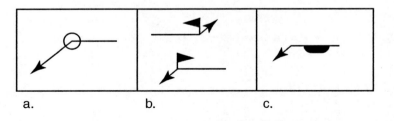

FIG. 30-2.

7. The type of weld indicated in Fig. 30-3 is ____.
   a. fillet   b. double fillet   c. seam
   d. plug   e. double slot

7. _____

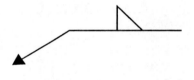

FIG. 30-3.

8. The type of groove indicated in Fig. 30-4 is ____.
   a. U   b. J   c. double-U
   d. double-J   e. square

8. _____

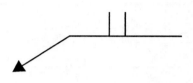

FIG. 30-4.

152 | STUDENT WORKBOOK

NAME: _____ DATE: _____

9. The welding should be done arrow side. (see Fig. 30-5)   9. _____
   (True or False)

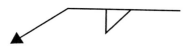

FIG. 30-5.

10. The welding on the ____ side are tack welds. (see Fig. 30-6)   10. _____

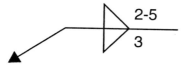

FIG. 30-6.

11. The distance from center to center of the tack welds is   11. _____
    called the ____.

12. The depth of the groove, arrow-side weld is ____. (see   12. _____
    Fig 30-7)
    a. 1/2"   b.   c.   2 1/2"   d. 2"   e. 6"

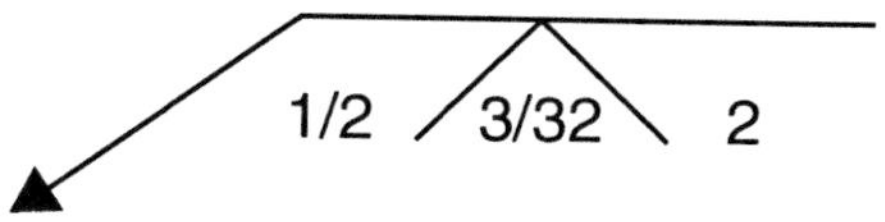

FIG. 30-7.

CHAPTER 30: Welding Symbols | 153

13. The size of the fillet weld, other-side weld is ____. (see Fig. 30-8)
    a. 3"   b. 5"   c. 3/8"   d. 1/4"   e. 2"

13. _____

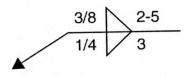

FIG. 30-8.

14. The size of the fillet weld, arrow-side weld is ____. (see Fig. 30-9)
    a. 3"   b. 5"   c. 3/8"   d. 1/4"   e. 2"

14. _____

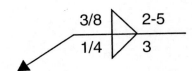

FIG. 30-9.

15. Multiple reference lines should not be used when a particular order in welding sequence is needed. (True or False)

15. _____

16. The included angle of the groove weld is ____ degrees. (see Fig. 30-10)

16. _____

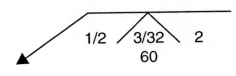

FIG. 30-10.

154 | STUDENT WORKBOOK

NAME: _____ DATE: _____

17. The black flag indicates a ____ weld. (see Fig. 30-11)   17. _____

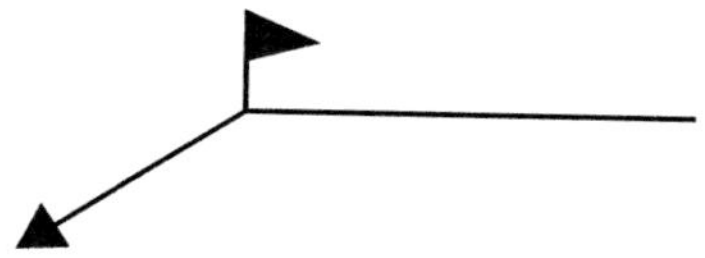

FIG. 30-11.

18. Which of the following is not a Nondestructive   18. _____
    Examination letter designation?
    a. LT   b. RT   c. PT   d. MT   e. NRC

                                                       Score _____
                                                            (26 possible)

CHAPTER 30: Welding Symbols | 155

NAME: _____ DATE: _____

CHAPTER 31

Safety

Choose the letter of the correct answer. If there is no lettered answer, write the word(s) answer.

1. A.c. transformer welding machines may be attached to a.c. lighting circuits. (True or False)

2. The ground lead to the work is adequate grounding for the welding machine. (True or False)

3. A welding machine may be moved safely while it is connected to the power source. (True or False)

4. When several lengths of cable are coupled together, connectors on both ground and electrode lines should be insulated. (True or False)

5. Size 2 copper cable may be safely used with 200 amperes of welding current with a 60% duty cycle welding machine. (True or False)

6. A container of water should be available for cooling hot electrode holders. (True or False)

7. The power supply must be disconnected for changing TIG electrodes or threading MIG equipment. (True or False)

8. Industrial alcohol cleans commutators quickly and safely. (True or False)

9. After welding, the entire welding machine may be blown out with clean, dry, compressed air. (True or False)

10. Air filters may cause overheating. (True or False)

11. The infrared and ultraviolet rays produced by the electric arc may "sunburn" an unprotected welder. (True or False)

12. Anyone over four feet away from the electric arc is safe from any harmful effects of its rays. (True or False)

1. _____
2. _____
3. _____
4. _____
5. _____
6. _____
7. _____
8. _____
9. _____
10. _____
11. _____
12. _____

13. Before welding, a welder should use a systematic approach to make sure you have not forgotten any of your PPE. (True or False)

14. Helmets and goggles should be sterilized before being transferred from one operator to another. (True or False)

15. For better vision during weld cleaning, helmets and goggles may be raised if the power-driven brush has a hood guard. (True or False)

16. The simplest form of ear protection is the large device that fits directly in your ear canal. (True or False)

17. One of the most effective ways to protect you from the hazardous effects of welding fumes is to keep your head out of the fume plume. (True or False)

18. High temperature and humidity increase the hazards of electric shock. (True or False)

19. Cotton clothing ignites more easily than wool. (True or False)

20. Even the ears must be protected during overhead welding. (True or False)

21. The low voltages required for arc welding eliminate the danger of severe electric shock. (True or False)

22. Keep electrical cable splices or insulation repairs within 10 feet of the electrode holder. (True or False)

23. The mixture of flammable gases and air in oxyacetylene welding torches is highly explosive. (True or False)

24. Gas cylinders may be lifted with a crane only when secured to a cradle or platform. (True or False)

25. Empty cylinders require no closing or capping. (True or False)

26. Well-constructed cylinders may be stored outside in an open area. (True or False)

27. Store acetylene cylinders on their sides to keep acetone and packing material from settling to the bottom. (True or False)

13. _____

14. _____

15. _____

16. _____

17. _____

18. _____

19. _____

20. _____

21. _____

22. _____

23. _____

24. _____

25. _____

26. _____

27. _____

NAME: _____ DATE: _____

28. A regulating device must be installed between an acetylene cylinder and the torch. (True or False)

28. _____

29. Crack acetylene cylinders before attaching regulators. (True or False)

29. _____

30. Use a match to test for acetylene leak. (True or False)

30. _____

31. Oxygen does not burn. (True or False)

31. _____

32. Oxygen may be substituted for compressed air. (True or False)

32. _____

33. Valves on oxygen cylinders must remain partly closed for welding. (True or False)

33. _____

34. An oxygen hose may be attached to any torch hose connection. (True or False)

34. _____

35. For changing torches, gases must be shut off at the pressure regulators. (True or False)

35. _____

36. When oxyacetylene welding is finished, the flame is extinguished by closing both cylinder valves at the same time. (True or False)

36. _____

37. When welding is stopped for only a few minutes, it is all right to close only the torch valves. (True or False)

37. _____

38. If torch valves freeze or become hard to operate, remove and oil the valve stem. (True or False)

38. _____

39. Clean clogged nozzle holes with any sharp drill. (True or False)

39. _____

40. It is permissible to use one hose for several different gases, as long as it is well cleaned between uses. (True or False)

40. _____

41. Hose splices should be securely taped. (True or False)

41. _____

42. Thorough visual inspection should determine whether or not a section of hose is safe for use after a flashback. (True or False)

42. _____

43. A standard pressure regulator may be used for several different gases. (True or False)

43. _____

CHAPTER 31: Safety | 159

44. The adjusting screw on a regulator must be released before the gas is turned on. (True or False)

44. _____

45. A fire extinguisher should be part of a welder's standard equipment. (True or False)

45. _____

46. When welding inside a confined space, such as a boiler, set both tanks inside so they may be turned off quickly. (True or False)

46. _____

47. Concrete is a safe working surface. (True or False)

47. _____

48. Too much oxygen pressure can blow sparks out of the work area. (True or False)

48. _____

49. Touching the torch tip to the work may cause preignition and backfire. (True or False)

49. _____

50. Even when a torch is improperly handled, flashback rarely occurs. (True or False)

50. _____

51. Oxygen from a welding cylinder may also be used to aid in ventilation. (True or False)

51. _____

52. Wooden floors should be covered with noncombustible materials or dampened before welding is begun. (True or False)

52. _____

Score _____
(52 possible)

NAME: _____ DATE: _____

CHAPTER 32

Welding and Bonding of Plastics

Choose the letter of the correct answer. If there is no lettered answer, write the word(s) answer.

1. The synthetic polymers (polymeric materials) have been used in the world economy for over ____ years.

2. Plastics are useful because of its excellent properties, corrosion resistance, light weight and fatigue resistant. (True or False)

3. Colors should always be used to identify a type of plastic. (True or False)

4. Plastics have properties similar to aluminum in that they have different specific ____ and surface hardnesses.

5. For thermosetting plastics all but which one of the following are true?
   a. harden when heated
   b. are formed by chemical reaction
   c. cannot be changed either by reapplying heated or chemical reaction
   d. cannot be welded

6. For thermoplastics which of the following is not correct?
   a. are used for wall panels
   b. soften when heated
   c. can be machined
   d. can be welded

7. ____ plastics are transparent.

8. Regarding welding thermoplastics which one of the following is not correct?
   a. similar to gas-welding metals
   b. fast, but somewhat expensive
   c. permanent
   d. possible in all positions

1. _____

2. _____

3. _____

4. _____

5. _____

6. _____

7. _____

8. _____

9. Layout work on thermoplastics should be done with all but which one of the following?
   a. lead pencil   b. soapstone
   c. a china marker   d. a scribe

10. Thermoplastics should be preshrunk. (True or False)

11. Cutting thermoplastics with a ____ saw produces less heat than other cutting methods.

12. Light-gauge plastic sheets may be sheared without producing stress marks if they are cooled slightly. (True or False)

13. As with metals, plastics are welded by applying enough localized heat to produce ____ of the areas that are to be joined.

14. Methods of welding plastics include all but which one of the following?
   a. Hot-Gas   b. Infrared   c. Conduction
   d. Microwave   e. Spin

15. Stretching may be caused by all but which one of the following?
   a. leaning the welding rod away from the direction of welding
   b. too little pressure on the welding rod
   c. plastic residue on the shoe
   d. insufficient cooling between multipass welds

16. Reduce distortion by speed welding with ____ welding rods.

17. Because PVC is a poor heat conductor, choose backup materials that conduct heat well. (True or False)

18. Environmental stress cracking results from ____ attack that would not normally be a threat.

19. All plastic welding torches are electrically heated. (True or False)

20. Plastic welding torches may provide a heat range up to ____ degrees F.

21. Compressed air or ____ gas is used as the hot gas medium.

9. _____

10. _____

11. _____

12. _____

13. _____

14. _____

15. _____

16. _____

17. _____

18. _____

19. _____

20. _____

21. _____

NAME: _____ DATE: _____

22. When welding plastics with electric torches, turn the gas on first and off last. (True or False)

22. _____

23. When welding plastics with an electric torch, ____ the gas volume to increase the welding temperature.

23. _____

24. When welding with a gas-heated, plastics welding torch, reduce the volume of the welding gas or ____ the pressure of the heating gas to raise the welding temperature.

24. _____

25. At the end of the welding operation with the gas-heated, plastics welding torch, always turn off the heating flame before shutting off the welding gas. (True or False)

25. _____

26. Plastic welds that are equal to less than ____% of the base material strength are considered unsatisfactory.

26. _____

27. Welded joints are the sites of potential weakness in a plastic structure. (True or False)

27. _____

28. Match the following defects of plastic welds to their possible causes.
　1. Porosity　　　　　a. Too little root gap.
　2. Poor penetration　b. Shrinking of base material.
　3. Scorching　　　　c. Stretching the welding rod.
　4. Warping　　　　　d. Rod and base material of different composition.
　5. Stress cracking　　e. base material too cold.

28a. _____
28b. _____
28c. _____
28d. _____
28e. _____

29. Match the following plastic weld defects to the appropriate corrective steps.
　1. Scorching　　　　a. Weld rapidly.
　2. Distortion　　　　b. Use small rod at root, large rods at top of weld.
　3. Warping　　　　　c. Back up weld with metal.
　4. Poor appearance　d. Increase air flow.
　5. Poor fusion　　　e. Use proper rod angle.

29a. _____
29b. _____
29c. _____
29d. _____
29e. _____

30. Destructive tests for plastic welds include all but which one of the following?
　a. tensile　　b. bending　　c. chemical　　d. impact

30. _____

31. The most effective way to test welds on plastic pipe is to subject the fabrication to the ____ test.

31. _____

CHAPTER 32: Welding and Bonding of Plastics | 163

32. To detect pores and cracks not visible in plastic welds by any other type of inspection, use ____.
    a. tensile tests
    b. impact tests
    c. spark coil tests
    d. radiography

32. _____

33. When welding plastics you must do all but which one of the following?
    a. place large beads at the base of the weld
    b. be sure the rod remains round
    c. be careful not to char the rod or base material
    d. not use flammable gases

33. _____

34. The most efficient, but expensive, method of inspecting the internal characteristics of a plastic weld is ____.

34. _____

35. For welding plastics, all but which one of the following is true about the filler rod?
    a. should be the same length as the weld
    b. should be forced ahead of the weld to conserve materials
    c. must not be stretched
    d. must be of the same composition as the base material

35. _____

36. A tensile strength value of ____ percent of the base material is considered acceptable.

36. _____

37. Heat for welding plastics is supplied by a low-heat torch flame. (True or False)

37. _____

38. Heating gases for plastic welding may be all but which one of the following?
    a. compressed air
    b. nitrogen
    c. oxygen
    d. inert gas

38. _____

39. Decreasing the volume of welding gas for joining plastics ____ the temperature of the gas.

39. _____

40. Torch heating capacity may be altered by changing the heating element and ____.

40. _____

41. For tack welding plastic joints all but which one of the following are true?
    a. assembles parts quickly
    b. eliminates the need for jigs and fixtures
    c. contributes to good fusion
    d. ensures proper part alignment

41. _____

NAME: _____ DATE: _____

42. Plastics do not need to be clean and dry prior to welding and during the welding operation. (True or False)

42. _____

43. Successful beading on plastics depends upon the proper combination of ____ and heat.

43. _____

44. A slight yellowing of the filler rod and base material is caused by ____.

44. _____

45. One of the most critical parts of making a quality weld is maintaining the proper temperature. (True or False)

45. _____

46. For high speed welding of plastics all but which one of the following are true?
    a. is even faster with a 45-degree torch angle
    b. does not produce flowlines
    c. forms a higher crown than hand welding
    d. must keep moving to avoid rod softening

46. _____

47. When joining plastic sheet, the welding proceeds only in a ____ direction.

47. _____

48. ____ welding of plastic pipe produces a joint that is stronger than both the pipe and the fitting.

48. _____

49. For plastic pipe all but which one of the following are true?
    a. must never be stored in the sun
    b. must be carefully cut
    c. may be heated and bent
    d. may be installed in a cinder fill

49. _____

50. Adhesive Dispersion is commonly used in plastic joining. (True or False)

50. _____

Score _____
(58 possible)

CHAPTER 32: Welding and Bonding of Plastics | 165